The Big Splat, or How Our Moon Came to Be

The Big Splat, or How Our Moon Came to Be

Dana Mackenzie

WILEY

John Wiley & Sons, Inc.

Contents

Genesis Revisited

There are many ways of looking at the Moon: with awe, with reverence, with longing, with fear. It is at once familiar yet mysterious, distant yet near, constant yet ever changing. Sometimes it seems close enough to be part of the Earth, at other times it seems as remote as the cosmos. High in the sky, it plays cat and mouse with the clouds, ducking behind them and then peeking out as if they were a flimsy veil. At moonrise it looms on the horizon like a big orange mountain, dwarfing houses and trees. On a spring night it consorts with Venus among the stars, the evening star dangling like a diamond earring from the Moon's crescent ear.

Men and women also have looked at the Moon for millennia as a practical aid for life on Earth. It has served as a torch for travelers, a timekeeper for farmers, a location finder for mariners. Occasionally it was a harbinger of doom, blocking the Sun during a solar eclipse, or turning bloodred during a lunar eclipse. Even in today's world of precision chronometers, its old role as timekeeper shows up in almost every culture. Christians still use the Moon to set the date of Easter; Muslims break Ramadan when they sight the crescent Moon; and countries such as China and Vietnam still use a lunar calendar along with the Western one.

Only in the past four hundred years have we begun to look at the Moon through the eyes of modern experimental science. With the invention of the telescope in the seventeenth century, astronomers such as Galileo Galilei could, for the first time, summon the Moon closer and inspect its surface. A whole new Moon emerged, a world unto itself with mountains, "seas," and innumerable pockmarks that astronomers called craters because of their resemblance to volcanic craters. The Moon became, for the first time, a place with features one could name: the lunar Apennines, the crater Tycho, the Sea of Tranquility.

Not long after scientists began to conquer distances, they began to conquer time as well. They recognized that underground layers of rock are like pages in Earth's geological history. The fossils in the rock told of earlier epochs of life on Earth. In the heavens, they found nebulas and galaxies, protostars and perhaps proto-solar systems in the process of formation, and they wondered whether these could offer a glimpse into an earlier era of our own solar system. Inevitably, irresistibly, they were drawn to speculate on the origin of species, of stars and planets, and of our own Moon.

"Men will always aspire to peer into the remote past to the utmost of their power," wrote George Howard Darwin, "and the fact that their success or failure cannot appreciably influence their life on earth will never deter them from such endeavors." Darwin should have known: his father, Charles Darwin, was the great evolutionary theorist who wrote *The Origin of Species*. George Darwin is not as familiar a name, to modern readers, as Charles. In George's day, though, he was one of the leading scientists in England, and in 1905 he followed in his father's footsteps to be knighted by England's monarch.

Darwin followed in his father's footsteps in another way. Beginning in 1878, he developed what might be called an evolutionary theory of the Moon, although it is more commonly called the "fission theory." Darwin argued that the Moon could have split off from proto-Earth when it was still a liquid body, flung off by Earth's rapid rotation and the action of the Sun's tides. After that it gradually moved outward over the aeons to its present position. Darwin's theory, which he arrived at by applying accepted physical principles about the action of tides, was the first scientific speculation about the origin of the Moon that treated it as a unique event, rather than an unremarkable part of an ongoing process of the formation of the solar system.

For a while George Darwin's idea reigned supreme, but by the 1930s more careful calculations of the tidal effects had thrown it into doubt. Two more theories arose to challenge it: the "capture theory," according to which the Moon was formed independently of Earth and subsequently captured by Earth's gravity; and the "coaccretion theory," which said that the Moon and Earth had formed together out of essentially the same raw materials. With so many theories and

George Howard Darwin (1845–1912), son of the famous naturalist Charles Darwin, proposed the "fission" theory for the origin of the Moon. He was also the world's foremost expert on the theory of tides, and proposed a theory of tidal friction to account for the Moon's gradual movement away from Earth. *Photograph courtesy of Cambridge University Library.*

so little hope for deciding among them, the whole problem of where the Moon came from became a bit of a nuisance to scientists.

Then, in October 1957, the Soviet Union launched *Sputnik I*, the first artificial satellite. Fewer than four years later, in April 1961, U.S. president John F. Kennedy made his famous public commitment to send men to the Moon and back before the end of the decade. Suddenly scientists had a real opportunity to get some hard answers about the Moon, if they could only get on board. It was by no means a certainty, in the beginning, that any science at all would be done on the Moon. The Moon mission could have been nothing more than a public relations stunt, as suggested in a 1962 news parody in the *New Yorker*, in which the Soviets send an orchestra to the Moon while the Americans are still struggling to get a rocket into orbit.

Two people were most responsible for making sure that science got on the lunar agenda. Harold Urey, who had won the Nobel Prize in chemistry in 1934 for his discovery of deuterium (a heavy isotope of hydrogen), was the first big-name scientist on the bandwagon, advocating Moon studies even before NASA was formed. He was passionately interested in the Moon's origin, loosely committed to the capture hypothesis but ardently committed to the idea that the Moon had started out cold, not hot, as Darwin had assumed.

Eugene Shoemaker, unlike Urey, had his greatest moment of glory still ahead of him: he would be remembered as one of the codiscoverers of Comet Shoemaker-Levy, which plunged suicidally into Jupiter's atmosphere in 1994 and spectacularly confirmed, in living color, the reality of collisions in the solar system. He also is the only person buried on the Moon: his ashes were carried there by the *Lunar Prospector* spacecraft, which crash-landed in 1999.

Shoemaker was a geologist and, if the truth be told, had little to say about the origin of the Moon. But he did have very strong beliefs about the origin of the craters—another hotly debated question in pre-Apollo days. Shoemaker believed that the great majority of craters had been created by the impact of meteorites on the Moon's surface. It was really by pursuing Shoemaker's theory to its logical conclusion that William Hartmann, a self-described "crater counter," came up in 1974 with the story of the Moon's origin that most planetary scientists accept today. But for readers who like a modicum of suspense, I will leave that story for a later chapter.

On Christmas Eve 1968, humans for the first time orbited in the gravitational field of a planet that was not their own. On television screens, viewers around the world could see the cratered surface of the Moon up close, blurry but huge, seemingly close enough to touch. As the broadcast from *Apollo 8* ended, astronauts William Anders, Jim Lovell, and Frank Borman read from Scripture:

"'In the beginning God created the heaven and the earth,'" intoned Anders's staticky voice. "'And the earth was without form, and void, and darkness was upon the face of the deep. And the spirit of God moved upon the face of the waters, and God said, Let there be light: and there was light. And God saw the light, that it was good: and God divided the light from the darkness.'"

On the television screen, the Moon glided by in ghostly silence. It was easy to imagine the Supreme Being riding in the capsule with Anders, gliding over the face of the Moon just as in Genesis, dividing the light from the darkness of outer space.

Lovell and Borman took turns reading from the first book of the Bible, detailing the second and third days of creation. They stopped before the fourth, when the Sun and the Moon were created; perhaps they knew that this part of the story was due for a revision. "'And God saw that it was good,'" concluded Borman. "And from the crew of *Apollo 8,* we close with good night, good luck, a Merry Christmas, and God bless all of you—all of you on the good Earth."

Though some people later criticized the astronauts for reading from a religious text, their Christmas Eve broadcast was one of the emotional high points of the Apollo missions. It was the only time they fully lived up to the grandeur that was expected of them. No other moment, save perhaps the actual landing of *Apollo 11,* quite matched the poignance of those three distant voices reading those ancient but familiar words over an ancient and incomprehensibly alien landscape.

What better text for the occasion than Genesis? It was, after all, genesis that we were after. The trip to the Moon was a trip as far back in time as we can go, to a land older than any on Earth. The astronauts on later missions were trained to look for "genesis rocks." The very top question on Apollo's scientific agenda, a question that scientists had debated for nearly a hundred years, was to determine how the Moon got there in the first place.

The quest to understand the Moon's origin was the only scientific goal that could rival the audacity of going to the Moon in the first place. Of course, the clues would be indirect; no one was around to record the Moon's creation on television. Science proceeds by analogies and by reproducible experiments—but there is only one Moon, and no laboratory will ever be large enough to produce another. Moreover, planet formation (which includes moon formation) isn't exactly the province of any one science. Geologists can tell how a rock forms, but they can do so only if a context for the rock—a planet—already exists. Physicists can track a planet's orbit for billions of years, but they cannot say where it started from or what it was before it was a planet. Chemists can work out a planet's composition as surely as they can identify the compounds in a scrambled egg; but they can't unscramble the egg and describe how it was put together. Astronomers understand more or less how stars are put together, but even they have never witnessed the birth of a moon.

It's no surprise, then, that the answer emerged slowly—too slowly for a restless public to sit around and wait for it. By the time the story of the Moon's birth finally came out, the last lunar lander had long since been put on exhibit in a museum, the last Saturn booster turned into a truly immense lawn sculpture, and the last Moon rock locked away in the Lunar Receiving Laboratory. When it finally appeared, the story was tucked away in astronomical journals that the public never reads. Perhaps the wonder is that it ever came out at all. The trail could have been too cold, the clues too spotty or too contradictory.

There is another reason why the mystery took a long time to solve, even after we had all the clues. Scientists, like most people, prefer to look at the small picture: to analyze a particular rock, to measure a particular isotope ratio, to estimate a particular age. It takes a certain amount of courage to step beyond one's day-to-day experiments and look at the big picture—and the origin of the Moon is a "big picture" question *par excellence.* Perhaps it makes sense that William Hartmann, one of the two scientists who unraveled the Moon's biggest mystery, is not only a scientist but also a part-time artist and science-fiction writer. It took someone with an artist's eye and a fiction writer's speculative temperament to see the big picture.

This is a book about that big picture: the origin of the Moon, as interpreted by Hartmann and Alastair Cameron, the second patriarch of "The Big Splat." It is also about a doomed planet called Theia, and a familiar one called Earth that used to look vastly different from today's Earth. But most of all, it is about a long lineage of intellectual voyagers who began exploring the Moon long before Neil Armstrong planted his boot into the lunar dust. The lineage contains some household names (at least to scientists): Galileo, who first brought the Moon within arm's reach, and Johannes Kepler, who always believed that Copernican astronomy would take us to the Moon. It also includes some less familiar names, such as George Darwin, who may be unjustly neglected now because their theories are no longer in fashion. And it includes a bevy of scientists of today, who are still struggling to put together sketchy clues into a coherent history of our celestial companion. Even though Hartmann and Cameron have given us a framework, many of the details of the story are still uncertain.

As the text chosen by the *Apollo 8* astronauts shows, the Moon has extraordinary cultural resonance for all of us; our arts and traditions and religions have all been inspired by millennia of Moon-watching. I have not written very much in this book about the mythology of the Moon, because I have a story to tell about the Moon as seen by science. However, I do not want to minimize in any way the importance of the Moon as a cultural symbol. It is unfortunate, I think, that we *didn't* send an orchestra to the Moon—or an artist, a poet, a filmmaker, or anyone who could translate the spiritual meaning of what we had done and where we had gone. We need Moon enthusiasts as well as Moon scientists. It is to them—to everyone who has felt their breath catch as they looked out the window at our impossibly beautiful neighbor in space—that I dedicate this book.

1

A Highly Practical Stone

The Moon that Neil Armstrong and Buzz Aldrin walked on was already very different from the one our ancestors had worshiped. It was hard for people watching on television to feel a spiritual connection to this new Moon—seemingly a place of desolation, not of spirituality. Some of the astronauts, to their credit, tried to express the new and alien sense of grandeur they felt, and some came home deeply changed by their experience. But for those who watched at home—and those who didn't watch—the Moon had become just a place, and not a terribly inviting one at that.

But in truth, our disconnection with the Moon began long before 1969. It has been going on for centuries. The Moon used to be more than a religious symbol and more than just a pretty ball in the sky: it was an integral part of daily life. Farmers used it to guide them in planting crops, and sailors watched the tides it produced. Travelers used it to find their way at night, and everybody used it to keep track of the passage of time. But over the centuries, all of these practical uses for the Moon have become weakened or obsolete as people developed a more sophisticated understanding of the natural world.

"The Parish Lantern"

It is hard for us to imagine today how utterly different the world of night used to be from the daylit world. Of course, we still can re-create something of that lost mystique. When we sit around a campfire and tell ghost stories, our goose bumps (or our children's) remind us of the terrors that night used to hold. But it is all too easy for us to pile in the car at the end of our camping trip and return to the comfort of our incandescent, fluorescent, floodlit modern world. Two thousand years ago, or even two hundred, there was no such escape from the darkness. It was a physical presence that gripped the world from sunset until the cock's crow.

"As different as night and day," we say today. But in centuries past, night and day really *were* different. The social order was inverted at night. At night the robber and the hoodlum were king, while law-abiding citizens cowered at home and the night watchman patrolled uneasily by torchlight. Night was, of course, the time for supernatural terrors as well, ghosts and werewolves and fairies and witches. Even whistling at night was dangerous—you risked inviting the devil.

In a time when every scrap of light after sunset was desperately appreciated, when travelers would mark the road by piling up light stones or by stripping the bark off trees to expose the lighter wood underneath, the Moon was the traveler's greatest friend. It was known in folklore as "the parish lantern." It was steady, portable, and—unlike a torch—entailed no risk of fire. It would never blow out, although it could, of course, hide behind a cloud. According to the Bible, that was why God gave us the Moon. "And God made two great lights; the greater light to rule the day, and the lesser light to rule the night."

Nowadays we don't need the Moon to divide the light from the darkness because electric lights do it for us. Many of us never even see a truly dark sky. According to a recent study on light pollution, 97 percent of the U.S. population lives under a night sky at least as bright as it was on a half-moon night in ancient times. Many city-dwellers live their entire lives under the equivalent of a full moon.

Although it took Thomas Edison's invention of the electric light-bulb to sever our dependence on moonlight, its influence was waning even before that. As early as the 1660s, Louis XIV (the "Sun King") installed the first permanent streetlights in Paris, to discourage the bands of vandals who had until then roamed unchecked at night. In 1784, a Swiss physicist named Aimé Argand discovered that he could increase the brightness of an oil lamp tenfold by making the wick hollow, so that the flame could be fed by air from the inside as well as the outside. Argand lamps were the first artificial lights that could compete with the full Moon, but another invention soon outshone them. The gas lamp, invented in 1799 by Englishman William Murdoch, was ten times more powerful. Suddenly the night-time was transformed: shops could remain open after dark, and factories could run all night. "Noctambulism" became a new fad in Paris, which became the "City of Light."

In 1879, when Thomas Edison finally invented the electric light-bulb, some people even thought its brightness was overkill. "Such a light as this should shine only on murderers and public crime, or along the corridors of lunatic asylums, a horror to heighten horror," wrote the author Robert Louis Stevenson in 1893. "To look at it once is to fall in love with gas." Nevertheless, the public hungered for more light. The gaslight faded into technological oblivion, and the Moon, a hundred times fainter than Edison's light, now circled the sky in forgotten splendor.

Marking Time

A similar tale, though a much older one, can be told about the Moon's other great service to humanity and how we eventually outgrew it.

The Greek philosopher Plato expressed a different view of the Moon's purpose from the one in Genesis. "Time came into being with the heavens in order that, having come into being together, they should also be dissolved together if they are ever dissolved," he wrote in the dialogue *Timaeus*. "As a result of this plan and purpose of god for the birth of time, the sun and moon and five planets . . . came into being to define and preserve the measures of time."

Of the celestial bodies Plato mentioned, the Moon certainly was the first one that humans turned to for timekeeping for time scales longer than a day. The most celebrated prehistoric site in Europe, the cave paintings in Lascaux, France, include a probable lunar calendar that is fifteen thousand years old. Among the vibrant animal pictures in the Chamber of the Bulls, one also can find some curious patterns of dots. Some of these may represent constellations, such as the Pleiades. But others are too regular, such as a straight row of thirteen dots terminated by an empty square. What could thirteen dots in a row mean? According to Michael Rappenglueck, a historian of astronomy at the University of Munich, they might represent half of a lunar cycle: the time from the new moon to the full moon (or vice versa), with the empty square representing the day when the Moon disappears.

Deeper in the cave is an even more striking example: a dappled brown horse with a black mane, and a meandering row of black dots trailing from its muzzle. To the modern eye it looks as if the horse has broken free and is galloping along with its reins trailing behind it. But that naive interpretation is historically impossible: In the era of the Lascaux paintings, the horse was still a wild animal, and reins did not exist. Again, says Rappenglueck, the clue to the dots' meaning is their number. There are twenty-nine of them—this time, the number of days in a complete lunar cycle. "It was a rhythm of nature that was important to these people," says Rappenglueck. Evidently, the Moon's motions in the sky meant as much to them as the motions of their prey on the ground.

Although it takes a little bit of creative mind reading to discern a lunar calendar in a fifteen-thousand-year-old cave painting, there is no doubt that humans were attuned to the cycles of the Moon by the time the first civilizations emerged five thousand years ago. Throughout the world—in Egypt, Sumer, Central America, and Asia—the first calendars were lunar. In Asia and the Muslim world, lunar calendars remain in use to this day.

To appreciate the convenience of a lunar calendar, imagine yourself in a world with no wristwatches to keep track of the hours and days. Imagine no pinup calendars, no desk organizers with witty cartoons. Imagine no paper to print a calendar on, and no printing press to print it with. You have nothing to tell time with but your own wits. How might you do it?

If you insisted on using a solar calendar, you would have to count out 365 days from one winter solstice to the next. Counting to 365 is harder, of course, than counting to 29. Even if you kept count by making marks on a stick, or a cave wall, you would have that many more chances to make a mistake. Most likely you would entrust the matter to the village priest or medicine man. But even if he kept good records, it wouldn't be so easy for him to determine exactly when the beginning of the year is. The solstice is the time of year when "the Sun stands still"—that's what "solstice" means in Latin. In other words, the Sun's highest point in the sky that day is almost the same as its highest point the next day, and the day after that. The very thing that makes the day unique also makes it hard to identify. It's like trying to time the moment when the stock market hits its peak.

Of course, one might argue that the winter solstice is the shortest day of the year, so the medicine man would only need to measure which day has the shortest time between sunrise and sunset. But keep in mind that your medicine man would have no way to measure hours and minutes accurately. In the ancient world there were no stopwatches, nor even any absolute measurements of time. When the hour was invented by the ancient Egyptians, it was simply one-twelfth part of a day or one-twelfth part of a night. So every day was twelve hours long—there was no "shortest day"!

All right, then, why not erect some kind of very visible monument, so that when the tip of that monument's shadow touches a certain point, or when the Sun is just visible through a crack in the stone, you would know that it is midwinter or midsummer? That is exactly what several ancient cultures did, from the builders of Stonehenge to the carvers of petroglyphs in Mesoamerica. But this solution to the solstice problem requires, at a minimum, a high degree of social organization. Moreover, it is not very portable.

All in all, a solar calendar is beginning to seem like a very expensive and inconvenient proposition. Contrast this with the simplicity of a lunar calendar. Instead of counting to 365, you only need to count to 28 at most, the number of days between new Moons. It's much easier to remember that today is the 11th day since the new Moon than to remember that it's the 164th day since the winter solstice. And even if you forget, the Moon itself can remind you. It changes its shape visibly in the sky from night to night. With a little experience, you can easily tell an 11-day-old Moon from a 9-day-old one.

For all these reasons, many early societies were tied to the rhythm of the Moon. The days of the month were more than numbers on a page. Each day and each time of the month acquired its own meaning. Hesiod wrote a book called *Works and Days,* which might be called the *Farmer's Almanac* of ancient Greece. Along with all sorts of choice advice on matters large and small, from when to marry to which direction to face when urinating, Hesiod also tells which days of the month are propitious for which actions: "On the fourth of the month bring home your bride, but choose the omens that are best for this business. Avoid fifth days: they are unkindly and terrible. . . . On the great twentieth, in full day, a wise man should be born. Such a one is very soundwitted. The tenth is favorable for a male to be born; but, for a girl, the fourth day of the midmonth."

No explanation is given for most of these admonitions, but the Greeks must have taken them seriously, as Hesiod's work is one of the oldest texts preserved to modern times. We still can see an echo of this tradition in the nursery rhyme "Monday's child is fair of face,

Tuesday's child is full of grace." In ancient times these characteristics would have been ascribed to the days of the month, not the days of the week.

For cultures that were advanced enough to use either kind of calendar, the choice between lunar and solar calendars depended on the culture's priorities and the use it made of that calendar. The fact that the solar calendar is used in most of the Western world does not make it "better."

In Islam, each month begins when two reliable witnesses sight the crescent Moon. According to some traditions, the observers must be local, which means that the month may begin on different days in different parts of the world. (The first crescent may not yet be visible when the Sun sets in Tajikistan, but several hours later, when the Sun sets in New York, the crescent may be wide enough to become visible to the naked eye.) Because a physical observation of the Moon may depend on atmospheric conditions, the first day of the month cannot necessarily be predicted in advance. To a non-Muslim, this unpredictability might seem like a nuisance, but consider also the spiritual importance of this custom. Every Islamic community is in charge of its own time, and cannot rely on the learned pronouncement of some distant authority. Every community is joined intimately to the sky, and through the sky to Allah.

Another possible inconvenience of a lunar calendar stems from the fact that twelve lunar months add up to a little over 354 days. This means that the Islamic year is 11 days shorter than a solar year, and hence significant religious events such as Ramadan gradually migrate from season to season. Of course, Islamic astronomers could have produced solar calendars if they had wanted to; they were the world's most proficient astronomers until at least the European Renaissance. It simply wasn't important to them to synchronize their calendar to the seasons.

However, many other cultures have tried to devise calendars that synchronized both the lunar cycle and the solar cycle—a "luni-solar" calendar. They named moons after events that occurred at a specific time of year: harvest moon, hunter's moon, honey moon. But to make such a scheme work, it was necessary to add, or "intercalate," an extra moon in some years. Otherwise the harvest moon would slowly drift into spring.

An imperfect way to bring the lunar and solar cycles into harmony was instituted by the Babylonians as long ago as 528 B.C. Since

12 lunar cycles fall 11 days short of a solar year, it stands to reason that over an 8-year period 96 (8 × 12) moons would fall 88 (8 × 11) days short of 8 complete solar years. But 88 days is very close to three lunar cycles, so the lunar and solar calendars could be kept in step by adding three months at prescribed times every eight years.

A century later, an Athenian astronomer named Meton found an even better match: 19 years are almost exactly the same as 235 months. Using modern values for the length of the tropical year (365.2422 days) and the synodic month (29.53059 days), we can calculate just how good the match is. In 432 B.C. the Greeks adopted the Metonic cycle as their official calendar, with a regular program of intercalated months that repeated every 19 years. If we had stuck to this calendar, the year 2000 would have ended the 128th Metonic cycle, and the lunar and solar calendars would be only 11 days out of sync. We would not need to change the calendar until A.D. 6032, when the two calendars would be off by a month and we would have to skip one of the scheduled intercalary months. Not bad for a calendar devised before anyone knew that the Earth rotates around the Sun.

In parts of the world other than Babylon and Athens, the addition of months remained the province of emperors or priests. This system was not only haphazard but also easily abused: the authorities could prolong a year when it suited them, perhaps to collect more taxes or to start the new year on a more propitious day. Eventually, though, even those in power lost patience with this arbitrary and scandal-ridden system. Coincidentally, the next calendar reform happened at more or less the same time in both China and Rome.

In China, the Han emperor Wu Ti adopted a reformed luni-solar calendar in 104 B.C., devised by a court astronomer named Lo Hsia Hung. Lo did his job very well: it remained the official calendar for more than 2,000 years. Now called the "Farmer's Calendar" (Nung Li), it is still used to determine Chinese holidays, such as the New Year or the Autumn Moon Festival (the fifteenth day of the eighth moon). The Nung Li calendar still used intercalary months, and began each month with the new Moon, but the timing of the extra months was now based in an elaborate way on the Sun's position in the zodiac.

Fewer than sixty years later, the Western world also acquired a new calendar that it would use, with only one significant change, until the present day. In 46 B.C. Julius Caesar, fed up with a calendar that was out of sync with the Sun by 80 days, decreed that the following year (which became known as the "Year of Confusion") would have 445 days. Thereafter, a leap day—not a leap month—would be inserted every fourth year.

The Julian calendar is a full-fledged solar calendar. It was not the first, but it was certainly the most influential, thanks in part to the power of Rome and the military gifts of Caesar himself. In this calendar the months are, astronomically speaking, pure fictions. A new Moon can happen on any day of the month, unlike the Nung Li calendar, and there can even be two new Moons in a month. Since Julius Caesar, the Western world has been a place very different from Plato's world; the Moon is no longer our primary timekeeper.

The major refinement to the calendar since Caesar's time was a change only in accuracy, not in kind. This was the adoption of the Gregorian calendar, which took place in some countries in the sixteenth century, and elsewhere not until the twentieth. The Gregorian calendar is still solar, but reflects the fact that the solar year is about 365.2422 days, rather than 365.25. Thus one day needs to be removed roughly every century (subtracting .01 day from the length of the year), but added back once every 400 years (adding .0025 day). February 29, 2000, was for some countries the first time that the second correction was implemented.

In the last few years of the twentieth century, scientists also began to add leap seconds to the calendar, compensating not for the length of the *year* but the length of the *day*. The Earth's rotation is simply not as steady as the atomic clocks that are now used as the absolute standard for measuring time. In the space age, seconds matter.

Ironically, we have now returned to the pre-Caesar era, where a high priesthood (the keepers of our atomic clocks) decide more or less arbitrarily when it is time to insert a leap second. A second, of course, is such a trifle that most people never notice or care about the difference. But it is tempting to speculate that this is the first step to a calendar that is neither lunar nor solar—perhaps the "stardates" that are familiar to *Star Trek* fans? Perhaps, when this happens, we will cease to feel connected either to the Moon or the Sun, but will be connected to the cosmos instead.

In Plato's opinion, the Sun and the Moon gave us much more than a calendar. "The sight of day and night," he wrote, ". . . the months and returning years, the equinoxes and solstices, has caused the invention of number, given us the notion of time, and made us inquire into the nature of the universe; thence we have derived philosophy, the greatest gift the gods have ever given or will give to mortals."

Science itself, which falls under the broad umbrella of what Plato meant by "philosophy," grew out of the hard and tedious work of Meton, Lo Hsia Hung, and other ancient astronomers, patiently and with incredible accuracy recording their observations of solstices and equinoxes and eclipses. Throughout much of this book, the grand theorizers will occupy center stage, and the patient collectors of data will stand in the background. But science has always depended in equal parts on meticulous observation and speculation. As we shall see, where the Moon is concerned, hard facts have often been in shorter supply than entertaining theories.

2

The Stone Star

The Gallipoli Peninsula, a strip of land overlooking the waterway from the Mediterranean Sea to the Black Sea, has long been a site of epic encounters. In World War I, an abortive English campaign on this peninsula cost half a million lives and nearly ended the political career of Winston Churchill. In 401 B.C., the Peloponnesian War ended in the same place, when the Spartan general Lysander defeated the Athenian navy in the Battle of Aegospotami.

But a few decades before Lysander, this rural corner of Thrace, where goats outnumbered people, witnessed a cataclysm of a rather different sort. One day in 467 B.C., the earth was shaken by a thunderlike sonic boom, a puff of smoke, and a searing light. The goat herders found a strange black boulder, the size of a horse cart, that had apparently dropped from the blue sky. Word soon spread throughout Greece of "the stone star that fell down flaming at Aegospotami."

Of course, meteorites have fallen from the sky throughout history, even before recorded history. But this particular boulder fell at a most propitious place and time—the early years of classical Greece. It was a time when scholars were first beginning to look for rational explanations of the phenomena they saw in the skies. Not satisfied with being mere spectators at the play of the Sun, the Moon, and the stars, they passionately debated what was going on behind the scenes. What were Earth, the Moon, and the Sun made of? Was the great blackness of space an empty vacuum, or was it filled with air? Did the cosmos go on forever, or did it have an edge? What were these lights in the sky? What made them go in circles? Why did some of them (the planets) move differently from the others? Until the ancient Greeks, such questions were the stuff of fables and legends. But in the sixth century B.C., a school of philosophers arose—the *physici,* as Aristotle called them, because they studied the physical world—who believed that human reason could reveal the answers.

Mind over Matter

Anaxagoras, a philosopher who was born in 499 B.C. on the western coast of Ionia (now known as Turkey, but then part of Greater Greece), belonged to the fourth generation of Ionian *physici*. From a young age he had heard of the legendary Thales of Miletus, who in 585 B.C. had predicted a solar eclipse. Modern astronomers believe he merely made a fortuitous guess based on cycles that had already been known for centuries by the Babylonians. Lucky or not, this prediction cemented his reputation as a wise man and prepared the way for the philosophers who followed him.

Anaxagoras knew also the ideas of Thales' successor Anaximander, who argued that Earth was the center of all things and that the heavens were made of fire. The Sun and the Moon were not just disks but rings of fire that completely surround Earth. We cannot see the whole rings because they are separated from us by a veil of mist. But this veil is punctured by "breathing holes," like the nozzle of a bicycle tire, and through these breathing holes we can glimpse the fiery realm beyond. When one of the breathing holes gets covered up, he said, we get an eclipse of the Sun or of the Moon.

The third of the great Ionian *physici*, Anaximenes, had proposed instead that all things are made from air, which can come in denser or lighter forms. Thus a stone is a very dense form of air, while clouds are a very light one. In Anaximenes's system, the Sun and the Moon were the lightest eddies of air, which have drifted up like leaves in the wind and caught fire.

Anaxagoras was surely well versed in the theories of his three predecessors. But none of them had enjoyed the chance to look at an actual stone that had fallen from the sky. Anaxagoras, we may imagine, jumped on the first horse cart to Thrace to see this wonder with his own eyes. And when he returned to Athens (his adopted home), he lost no time in proclaiming his own theory: The Sun and the Moon were nothing but stones that had been flung off into space by the spinning Earth. The stars, too, were glowing stones. Every now and then one of them would come crashing back to Earth, like the meteorite at Aegospotami.

Anaxagoras (499 B.C.–ca. 427 B.C.), shown here in a somewhat fanciful German engraving from 1493, was the first Greek thinker to explain the origin of eclipses and to propose the idea that the Moon is a solid, Earth-like body. He believed that the Moon, like the Aegospotami meteorite and other "stone stars," had been flung off by Earth during its creation. *Photograph courtesy of Getty Images.*

However, Anaxagoras believed that the Moon and the Sun were very much larger than the meteorite—larger even than the Peloponnesus, the southern peninsula of Greece. Given his cultural frame of reference, the Peloponnesus was a reasonable comparison, because it must have been one of the largest geographical features known at that time. In essence, Anaxagoras was saying that the Moon was like a miniature Earth. As such, it must also have mountains and valleys.

Anaxagoras reached one other startlingly modern conclusion. Though solar eclipses had been observed for centuries, he was apparently the first person to deduce the reason why they occur. The Babylonians had known that eclipses tend to repeat themselves in a 19-year cycle. (This was the cycle that Thales presumably used to predict the eclipse of 585 B.C.) Certainly they knew that solar eclipses occur only at the new Moon. Yet somehow they had never made the connection that solar eclipses were *caused* by the Moon passing between the Sun and Earth. Perhaps they could not reconcile the appearance of the bright, shiny Moon at night with the dark, sinister creature that devoured the Sun in the middle of the day. Who can blame them? To the untrained eye, the Moon seems to be a giver of light, not a taker. And until Anaxagoras, no one had suspected that the Moon was a solid object that could block the rays of the Sun.

"Men of that day used to call [Anaxagoras] 'Nous' (The Mind)," wrote the biographer Plutarch, "either because they admired that comprehension of his . . . or because he was the first to enthrone in the universe, not Chance, nor yet Necessity, as the source of its orderly arrangement, but Mind pure and simple." Like the other "wise men" of his era, Anaxagoras eventually had his accomplishments exaggerated by those who did not understand them. Later historians recorded that he "predicted" the fall of the meteorite at Aegospotami, a feat that that would have outdone not only Thales but also all modern scientists, who have never yet predicted when a meteorite will fall on Earth and who probably will not do so for centuries to come.

Old beliefs and superstitions die hard, as Anaxagoras found out the hard way. Sometime after 450 B.C. he was brought to trial on a charge of impiety for teaching that the Moon and the Sun were stones and not gods. He was acquitted, apparently through the intervention of a friend in high places: his former disciple Pericles, who was now leader of the Athenian city-state. Nevertheless, Anaxagoras had to leave Athens and return to his homeland. As feisty as ever, he continued teaching in Ionia until his death in about 427 B.C. After his death, his townsmen erected an altar to Mind and Truth in the town square.

It is tempting to read Anaxagoras's trial romantically as an early skirmish between reason and prejudice, or between science and religion. But his prosecutors were undoubtedly motivated by politics as well as principle. Pericles, as the de facto ruler of what was supposed to be a democracy, had plenty of detractors. He was criticized, for example, for appropriating the religious offerings of other city-states to build magnificent temples in Athens. But it always has been easier to bring down the leader's associates than the leader himself, as presidents Reagan and Clinton have shown in our own day. Thus it was Anaxagoras who was put on trial, and he was not the only teacher of

Pericles to be hauled before a jury; Damon, his music teacher, suffered the same fate.

From the modern point of view, Anaxagoras was very wrong on some points. Like his three predecessors in the Ionian school, he believed in a flat Earth, which he placed at the center of the cosmos. His belief that the Moon was a stone turned out to be right, but we now know that the Sun is something quite different. Also, Anaxagoras's notions of how Mind made the universe go around were vague, to say the least. "Anaxagoras avails himself of Mind as an artificial device for producing order, and drags it in whenever he is at a loss to explain some necessary result," was Aristotle's tart critique.

Nevertheless, Anaxagoras got some things amazingly right. First and foremost, he made it possible to think of the Sun, the Moon, and the stars as solid, physical objects—a tradition that later Greek philosophers took up enthusiastically. His idea that the Moon spun off from Earth could be considered the first articulation of the "fission theory," which, as we shall see, George Darwin revived for very different reasons in the nineteenth century. Anaxagoras's intuition of a connection between meteorites and the Moon anticipated the work of Harold Urey and Gene Shoemaker in the twentieth century. His explanation of eclipses was right on the mark.

After Anaxagoras, Greek cosmology had two different schools of thought about the nature of the Moon and other heavenly bodies. Some philosophers were horrified by the idea that the heavens could be polluted by earthly substance, and continued to believe that the Sun and the Moon were divine beings. The latter camp included some of the most revered names of Greek philosophy: Socrates, Plato, and Aristotle.

Other philosophers followed Anaxagoras in viewing the Moon as a miniature Earth, and even broached the idea of extraterrestrial life. In the fifth century the followers of Pythagoras adopted this theory; it also fit in nicely with the views of the atomists, who believed that there were infinitely many worlds. It must have been a lively and exciting debate. Unfortunately, the details are hard to reconstruct now, more than two millennia later, because—as usual—the winners get to write history. The works of Aristotle and Plato have been preserved in massive tomes, studied and memorized and commented on for centuries. The writings and teachings of Anaxagoras and the rest of the "miniature Earth" camp have been preserved only in the barest of scraps, in legends and oblique allusions and offhand remarks by other writers. However, there is ample evidence that the victory of the "heavenly being" school was not complete or immediate, and that both possibilities remained in the public consciousness for several centuries.

From Pythagoras to Aristarchus

One place where Anaxagoras's view received a warm welcome was the city of Croton in Italy, where the legendary mathematician Pythagoras had settled in about 530 B.C. after fleeing Ionia. There Pythagoras had set up a tightly knit cult that eventually grew to include about three hundred people. With its mysticism and its strict rules for living, including vegetarianism, the cult would not have been out of place in New Age circles today. But its most unique aspect was the master's passionate belief in mathematics, which he saw as the key to understanding the universe; it was both a science and a form of worship.

Nowadays, high-school students first encounter the name of Pythagoras in connection with the Pythagorean theorem in geometry. But this seems like a paltry recognition of his influence. He practically invented the word "mathematics"—in the Pythagorean brotherhood, initiates were called *mathematikoi*. He was the matchmaker who wedded science to mathematics, a marriage that has lasted for twenty-five hundred years. It has been very much a love-hate relationship: many scientists over the years have made great discoveries by mathematical reasoning, but pure mathematics—especially if unsupported by physical evidence—also has led to a few fiascos.

The Pythagorean brotherhood produced many insights of lasting importance in astronomy as well as mathematics. First, the Pythagorean disciple Parmenides of Elea replaced the flat Earth of the Ionians with a spherical Earth. Besides the reasons frequently cited for this conclusion—the way that ships sink below the horizon as they sail into the distance, and the fact that the stars appear at different heights in the sky to observers at different latitudes—the Pythagoreans must surely have liked the beauty and geometric symmetry of the sphere.

Parmenides also argued that the Moon shone by reflecting the Sun's light. Another Pythagorean, Philolaus of Croton, went further and proposed that the Moon has a day and night, just like Earth. He was the first to realize that the lunar *day* is the same as a terrestrial *month*. Each time the Moon goes around Earth, it also makes a complete revolution with respect to the Sun; hence there is no such thing as the "dark side of the Moon," because every part of the Moon faces the Sun at some time during the lunar day.

Philolaus also believed that the Moon had animals and plants that were fifteen times larger than those on Earth. Presumably his reason was that if the lunar days were longer, the size of everything else had to increase in proportion. This may be cited as an early example of numerical reasoning gone awry; to a modern scientist, there are many factors in determining the size of a plant or an animal that are much more important than the length of a day (availability of water, for example).

By imagining life on the Moon, Philolaus was clearly taking a step away from the view of Earth as the center of the universe. His model of the solar system makes this explicit. He argued that the Sun, the Moon, and all the planets—*including Earth*—are in motion about a "central fire" called Hestia, the hearth of the universe. We can never see the central fire directly because a "counter-Earth" constantly stays between us and it. The Earth and its shadowy companion zip around the central fire once every day, thus accounting for the fact that the stars appear to circle the sky once every twenty-four hours.

However, an invisible planet *and* an invisible fire were too much for many people to swallow, and Philolaus's theory seems not to have convinced anyone outside the Pythagorean brotherhood. In the fourth century B.C., Heracleides of Pontus suggested a neater solution: Instead of *revolving* around a central fire, perhaps Earth simply *rotated* about its own axis. He was perhaps the first astronomer to distinguish these two different kinds of motion. Contradicting Philolaus, he reinstated Earth at the center of the universe. It took another century and another astronomer, Aristarchus of Samos, to put the ideas of Heracleides and Philolaus together, making Earth *both* rotate *and* revolve. The result was a view of the solar system that strongly resembles the Copernican theory, which would shock Europe more than seventeen hundred years later. Aristarchus wrote that Earth and all the other planets orbit the Sun, which does not move; Earth also rotates about its own axis; and this axis is at an oblique angle to the ecliptic (the plane of Earth's orbit).

The first part of Aristarchus's theory explains the Sun's apparent yearly motion around the zodiac—really, we are the ones moving, and the background stars we see behind the Sun change as we go around. Aristarchus's theory also provided a simple explanation for the other planets' more meandering motions, which will be discussed below. The second part, the rotation of Earth, accounts for the apparent daily motion of the stars in circles around the North Star. The third part accounts for the fact that the ecliptic—the circle of constellations through which the Sun and planets appear to move—is tilted with respect to the equator. It also accounts for Earth's seasons, although it's doubtful that Aristarchus realized that fact.

It was an amazing feat to disentangle the daily motions of the Sun, the Moon, and planets from their long-term motions, to recognize that Earth itself was moving, and to come up with a coherent vision of the solar system that explained so many observed facts. Aristarchus deserves to be just as famous for his heliocentric system as Copernicus is. But his theory did not just spring out of nowhere. It was the culmination of a natural sequence of ideas, leading from the Moon as an earthly body to Earth as a planetary body, orbiting the Sun. Certainly the earthly Moon fit the heliocentric theory much better than the more widely accepted geocentric theory advocated by Aristotle.

The Aristotelian Cosmos

Aristotle of Stagira, the son of the court physician to the king of Macedonia, studied for twenty years in Plato's Academy. He was widely considered Plato's prize pupil, but at Plato's death the leadership of the Academy fell to a nephew of Plato instead. Rather than remain in the background, Aristotle founded a rival institution of higher learning, the Lyceum. Like Anaxagoras, he had both the fortune and the misfortune to instruct a famous pupil: Alexander the Great. During Alexander's lifetime, Aristotle's reputation as the teacher of the world's greatest emperor must have stood him in good stead. But after Alexander died in 323 B.C., murmurs of "impiety" began to be heard against Aristotle, as they had against Anaxagoras and Socrates. Aristotle promptly left Athens so that, as he said, "The Athenians might not have another opportunity of sinning against philosophy."

Aristotle inherited from Plato the belief that everything in the universe was placed there for a purpose. But where Plato's writings tended to grow fuzzy around the edges when they touched on hard science and mathematics, Aristotle was an acute observer and an unequaled master of logical reasoning. Reading his works is a strange and unsettling experience for a student of modern science, because his theories are so much at odds with what we now learn and teach. Yet his arguments are presented with great logical clarity, exceeding the standards of many modern scientists. A typical Aristotelian passage runs like this. He presents a proposition he wishes to prove, such as: The heavens are made of a "quintessence" unlike earthly matter. He itemizes every possible opposing theory. He refutes, often with sublime arrogance, each one of them: "This need not trouble us, because . . ." And then he demonstrates the validity of his own position, often with not one but three or four different arguments.

Aristotle expounded his system of the cosmos most clearly in a book called *On the Heavens.* He accepted the existing theory of Empedocles that everything within the Moon's orbit is made of earth, air, fire, and water. He noted that each of these elements has a "natural" tendency to move either up or down. Heavier elements, such as earth, move down (toward the center of the universe), while lighter ones, such as fire, move up (toward the periphery of the universe). In either case, he said, they always move in a straight line unless they are forced to act differently by an "unnatural" force.

The Moon, the stars, and the planets, Aristotle said, also have a natural motion, but their natural motion is circular. This is in their nature because they are eternal and unchanging, and circular motion alone has no beginning and no end. By contrast, movement in a straight line always has a beginning and an end, and thus is natural for earthly objects, which cannot last forever.

Because the Moon and everything beyond it moved differently from objects on Earth, Aristotle theorized that they must be made of a different substance, which he called *aither.* (Later scholars would call it the fifth essence or "quintessence.") Thus Aristotle not only placed Earth at the center of the universe (where it had to be, because that is where earthly matter naturally congregates), he also placed an unbreakable qualitative barrier between Earth and the heavens. Earth was the realm of everything corruptible, temporary, changeable. From the Moon outward, everything in the universe was incorruptible, eternal, and pure. There could be no question, then, about the origin of the Moon or any of the other celestial bodies. They did not need to be created by a god, as Plato had said, or flung off from

Earth, as Anaxagoras had suggested. In Aristotle's system, the Moon was simply there, timeless and immutable, as it always had been and always would be.

However, the Moon was always a problem child for Aristotle, and even more so for his followers, because it is the only celestial body that is close enough to show details of its surface to naked-eye observers. Anyone who looked at the Moon could see dark spots on it, which were visible signs of its "impurity" or imperfection. If the heavens were changeless, why did the Moon go through phases? If the Moon were made of the same quintessence as the stars, why could it not shine with its own light instead of being forced to borrow the light of the Sun?

Aristotle resorted to a convoluted explanation. Apparently there were different levels of perfection. The stars, the most distant from Earth, were the purest of the pure. The planets, closer to Earth, were somewhat less perfect. This fact manifested itself in their motion, which had some very obvious deviations from "perfect" circularity, wandering north and south and sometimes backward in the sky. According to Aristotle they aspired to perfection, and therefore had motions that were combinations of several circular motions. Finally, the Moon, closest to Earth, was so far from perfection that it did not even aspire any more, but followed a circular path around Earth because that was its nature—like an animal following its base instinct.

Thus the Moon became a very important test case for philosophers and their competing cosmologies. It was on the border between the celestial and the mundane, and no one knew quite where to put it. Fortunately, a fascinating snapshot of the debate has been preserved to the present day. In about A.D. 75, the popular biographer Plutarch wrote *On the Face Which Appears in the Orb of the Moon,* a short treatise that shows what educated people thought about the Moon four centuries after Aristarchus and three centuries after Aristotle. Perhaps surprisingly, in view of his authority among ancient and later among medieval thinkers, Aristotle emerges as a rather badly beaten loser in Plutarch's account.

The First Popularizer

In studying the history of lunar science, one runs into two kinds of people. The majority are scholars who were working on larger questions that just happened to have implications for the Moon, more or less as an afterthought. The second, and smaller, category are ones who genuinely cared about the Moon for its own sake. Perhaps we can call them the "moonstruck" ones, or the "loonies"—in the friendliest possible way, of course. Most of our understanding of the Moon has come from people of the first stripe. As we have just seen, questions about the Moon tend to morph into questions about the solar system and vice versa, so there is no point in trying to build a wall between the two subjects. At the same time, I will try to point out the "loonies" when we encounter them. They are, after all, the heroes of this tale.

In this sense, Plutarch qualifies as our first loony. How else to explain the fact that a writer most noted for his biographies of famous men would compose a dissertation on the question of what the Moon is made of? And it is definitely the Moon that occupies the center of the stage in his book.

On the Face (*De Facie* in Latin) is a dialogue between several thinly disguised real people, although "dialogue" may be too genteel a word. It is an extraordinarily rich intellectual free-for-all. The narrator, Lamprias, defends the view passed down from Anaxagoras to Aristarchus, that the Moon is made of solid matter. It is clearly this position that Plutarch sympathizes with and gives the most time to. Lamprias's main opponents are "Aristotle" (not *the* Aristotle, but an Aristotelian whom Plutarch does not even bother to make up a name for) and Pharnaces, who argues that the Moon is made of fire. Interestingly, the main debate is between Lamprias and Pharnaces, with "Aristotle" playing a relatively minor role. Could this reflect the actual weight that these three theories had in Plutarch's time?

The dialogue strays far from the subject in the title, but it does begin with a discussion of the nature of the Moon's dark spots. Plutarch addresses several theories: (1) the spots are an optical illusion; (2) they are a mirrorlike reflection of Earth's seas; (3) they are places of "shadowy air" in the Moon's fire; or (4) they are shadows of mountains on the Moon. His narrator, Lamprias, dismisses the optical illusion theory right away, because the Moon is not bright enough to "bedazzle" the eyes and because sharp-eyed people can see the spots more clearly than people with "dull and weak eyes."

The mirror-reflection idea—which Plutarch attributes to a follower of Aristotle named Clearchus—is much more interesting. First, Lamprias says, the dark spots cannot be a reflection of Earth's seas because they are separated, while Earth's oceans are all joined. Second, the Moon cannot be reflecting Earth's image because Earth is not even in the center of the Moon's orbit! Here Lamprias is skillfully using one of the Aristotelians' own ideas against them; it was a mathematician named Hipparchus who had realized that Earth is offset from the center of the Moon's orbit. (The discrepancy is about two Earth diameters.)

Lamprias's argument—that we don't see our reflection in the Moon because it isn't really facing us—is ingenious but not really convincing. It would make sense only if the Moon were a flat disk. Later Plutarch brings a more compelling argument against the Aristotelian idea of the Moon as a surface with mirrorlike smoothness. If it were so, the Sun would be reflected in it as a point of light; instead, the face of the Moon is diffusely lit. As Lamprias puts it, "Reflections and diffusions of light from one another are multifariously reflected and intertwined and the refulgence itself combines with itself, coming to us, as it were, from many mirrors." He is right, although a modern-day scientist would say it more concisely: the light from the Sun is scattered from the Moon's surface.

Next, Lamprias and Pharnaces spar over the idea of the Moon as a fire. Pharnaces' best evidence seems to be that during a lunar eclipse, when the sunlight is blocked by the Earth and the Moon is presumably shining with its own light, it "displays a color smoldering and grim, which is peculiar to her." It is clear that Plutarch had no clue of the modern explanation for the Moon's red color during an eclipse (the diffraction of red light around the Earth, which also produces the red color of sunsets). Instead, Lamprias rebuts Pharnaces with a somewhat less convincing argument: If the Moon were made of fire, it would still need some kind of solid fuel to sustain the fire, and it would surely have run out by now. Also, he argues that if the Moon were made of fire, then at half Moon we would surely see some of that fire on the unilluminated portion, so that the Moon would appear white and red, rather than white and black.

Plutarch's discussion of the fourth theory, that the spots are shadows of mountains or valleys, is an interesting combination of quantitative reasoning and wishful thinking. A character named Apollonides, identified as a mathematician, points out that the shadows must be extremely large. True to his profession, he even makes an attempt to calculate how large they are. The Moon appears to be twelve finger breadths wide, and the spots half a finger breadth, so they are one twenty-fourth the diameter of the Moon. But how big is that? Aristarchus had already pointed out that the Moon is smaller than Earth, because it is completely covered by Earth's shadow during a lunar eclipse, and the shadow must be smaller than Earth. In fact, because the Moon takes some time to pass through the shadow, Aristarchus estimated that Earth's shadow—and therefore Earth—were at least twice the size of the Moon. By a geometric argument based on the estimated size and distance of the Sun, Aristarchus concluded that Earth's diameter was between 108/43 (2.51) and 60/19 (3.16) times the diameter of the Moon. The actual ratio is closer to 3.70, but certainly Aristarchus was in the right ballpark.

Since Earth's diameter is about 8,000 miles, Plutarch could therefore have worked out that the spots are at least 100 miles wide (8,000 divided by 3.16 times 1/24 [0.04]). But he fudged the figures a bit to get a smaller number; perhaps he was afraid that people wouldn't believe that a shadow of a mountain could be that wide. Actually, the problem with the "shadow" theory is that the Moon's spots can still be seen at full moon, when the Sun should be directly overhead and the shadows should be imperceptible. Plutarch gets around this with a truly far-fetched argument: "The reason is that the remoteness of the light alone and not the magnitude of the irregularities on the surface of the Moon has made the shadow large." In other words, the Sun is so far away from the Moon that it casts larger shadows there. Plutarch should have known that the Moon is not much farther from the Sun than Earth is, and also he should have known that the intensity of a light has nothing to do with the size of the shadows it casts.

Before judging Plutarch too harshly, though, we should remember that he was not a mathematician nor an astronomer but a writer. In *De Facie* he is, for the most part, not inventing his own arguments but repeating and organizing the arguments of other people, and he does a remarkably good job at that. He is especially effective when he rebuts the Aristotelian arguments against a solid Moon. Wouldn't a solid Moon fall to Earth? No, he says: "The Moon is saved from falling by its very motion and the rapidity of its revolution, just as missiles placed in slings are kept from falling by being whirled around in a circle." Wouldn't a solid Moon violate the Aristotelian doctrine of natural motion, which makes all earthly matter move toward the center of the cosmos? No, he says, even if one grants the idea of natural motion: "Let us calmly observe without any histrionics that this indicates not that the Moon is not earth but that she is earth in an 'unnatural' location. . . . If everything is always in its 'natural' position, I cannot make out what use there is of providence or of what Zeus . . . is maker and father-creator."

Written in the first century A.D., Plutarch's book gives us our last peek at how the west European world viewed the Moon, before the discussion was suspended by the fall of the Roman Empire and the loss of much of classical culture. We see the emergence of two themes: the Moon as something familiar, a chunk of earth, and the Moon as something alien, a part of the cosmos. These two ideas have coexisted uneasily in lunar studies to the present day. We see a culture that already possessed many relevant facts about the Moon and the solar system, even as it was handicapped by lack of scientific instruments (remember Apollonides measuring the spots on the Moon by finger breadths). And we see that the question of the Moon's nature (let alone its origin) was still very much open to debate. A scholar with Plutarch's reputation could still argue vigorously against the views of Aristotle that later generations accepted as dogma.

3

Kepler Laughed

On the Ides of March in 1610, a carriage clattered down the streets of Prague to the residence of the Imperial Mathematicus to Holy Roman Emperor Rudolph II. The mathematician in question, Johannes Kepler, was no stranger to court business; one of his main jobs was casting horoscopes for the wealthy and powerful. Nor was he a stranger to the occupant of the carriage. Privy Councillor Johann Matthaeus Wackher von Wackenfels, besides being a distant relative of Kepler, also was a close friend. The two often enjoyed debating matters both theological and astronomical.

There was much for them to talk about in those days. The aftershocks were just beginning to be felt from Nicolaus Copernicus's book *De Revolutionibus*, published in 1543, which said that the Sun was stationary and that Earth moved around it. Then there was Giordano Bruno's astonishing assertion, in the 1580s, that the stars were actually suns, and that there were infinitely many inhabited worlds just like our own orbiting them. Though Kepler was a dyed-in-the-wool Copernican, he thought Bruno went too far—Earth may not be the center of the universe, but the Sun certainly was. Not only were there new theories to argue about, but also strange new sights in the sky: new stars, or "novas," had appeared in 1572 and in 1604. What did they mean?

The news that Wackher was racing across Prague to convey would astound the world even more than any of this. The new stars, after all, had come and gone; some people thought they were not even stars at all, but an atmospheric phenomenon. But now, according to a rumor Wackher had heard in the imperial court, an Italian mathematician named Galileo Galilei had used a new instrument called a perspicillum (soon to be renamed a "telescope") to reveal four previously unknown *planets*. If it was true, it would be the most sensational astronomical discovery of the century—perhaps of all time. Where Galileo had found the new planets, the rumor didn't say.

Nevertheless, Wackher had his suspicions, and that was the second reason for his haste. He believed that Galileo had found the planets orbiting another star, which would settle one of his and Kepler's pet debates. For Wackher had defended Giordano Bruno's belief in other solar systems—a rather risqué position for a Catholic like Wackher to take, since the Roman Inquisition had burned Bruno at the stake only ten years earlier. If Galileo had really seen four planets orbiting another star, Kepler would have to admit that Wackher—and Bruno—were right.

By the time he got to Kepler's place, Wackher was ready to burst. We know what happened next from Kepler himself. Not even bothering to get out of the wagon, Wackher "told me the story from his carriage in front of my house. Intense astonishment seized me as I weighed this very strange pronouncement," Kepler wrote. "He was so overcome with joy by the news, I with shame, both of us with laughter, that he scarcely managed to talk, and I to listen."

For Wackher—who was a lawyer, not a scientist—it must have been a rare opportunity to score a point off of his world-famous friend. Yet how revealing it is, and how wonderful, that Kepler laughed, too. It was the laughter of astonishment, the laughter of being present at the beginning of something new and unforeseen. Kepler surely knew that such a strong instrument as Galileo's telescope would open up a new window on the universe. It would expose for good the Aristotelian hoax that had dominated astronomy for fifteen hundred years, and which Kepler himself had been fighting for nearly twenty years. Yet for Kepler, who was a deeply religious person, more was at stake than personal glory. The new instrument was nothing less than a new way of seeing God's creation. Something to laugh about, indeed!

Kepler's Early Career

Johannes Kepler was an unlikely astronomer: an illness in his youth had weakened his eyes and left him with multiple vision. But his contribution to the subject came mostly through his ideas, his mathematical talents, and a great deal of persistence.

Born in 1571 in the small town of Wyl (now Weil der Stadt), Germany, Kepler was a child of an absentee father—a mercenary who

Johannes Kepler (1571–1630) was one of the most ardent supporters of the new Copernican astronomy. Kepler's *Somnium (The Dream)* was the first attempt to extrapolate from scientific observations what life would be like on the Moon. *Photograph courtesy of Library of Congress.*

found many excuses to be away from home, and eventually vanished for good. At age seventeen, Kepler went off to the Lutheran university at Tübingen to study theology. Almost immediately he met a teacher who changed the course of his life. Michael Mästlin, the professor of astronomy at Tübingen, was one of the first scholars on the Continent to embrace the new theory of Copernicus. Martin Luther himself had condemned the theory, so we can only assume that Mästlin took some personal risk in teaching it, and perhaps he saved it for those students who seemed receptive. Kepler was more than receptive. From 1591 on, he dedicated his life to promoting and improving on the heliocentric theory.

If the Copernican theory was Kepler's passion, the Moon became, in some ways, his nemesis. The Moon enters his life story for the first time in 1593, when he and a younger student named Christoph Besold proposed to hold a "philosophical disputation" at Tübingen on the subject of the Moon's motion. From Kepler's later writings, it is pretty clear what his motivation was. According to historian Edward Rosen, "A common objection to Copernicanism was the argument that if the Earth really does move, it should provide its inhabitants with some perceptual evidence of its motion. The refutation of this common anti-Copernican argument was the principal purpose of the Kepler-Besold lunar thesis of 1593." In particular, Kepler hoped to argue that Moon-dwellers would assume the Moon to be stationary, because they cannot perceive its motion any more than we can perceive Earth's. Yet we know that, contrary to the Moon-dwellers' perceptions, the Moon does move. So why should the same not be true for Earth?

Unfortunately, the proposed public debate never took place, because the professor of philosophy in charge of disputations refused to allow it. In fact, it took forty-one more years for Kepler's explosive "lunar thesis" to come out in public—and it did so only after being greatly transformed, and only after it had nearly ruined his life.

In spite of his advocacy of Copernicus's theory, and in spite of the religious intolerance that was becoming more and more acute in Germany and Austria, Kepler's career progressed rapidly. In 1594 he received a teaching post in mathematics at a seminary in Graz, and the following year he made what he always considered to be the greatest discovery of his life.

While teaching a class one day in July 1595, Kepler suddenly hit upon the reason why, as he saw it, God had created six planets—Mercury, Venus, Earth, Mars, Jupiter, and Saturn, the only planets known at that time. Mathematicians had known for centuries that there are only five "Platonic solids." These are polyhedra—or, if you prefer a less technical word, faceted solids—in which every face is identical, and every face is a regular polygon (either a triangle, a square, a regular pentagon, or a regular hexagon). These five Platonic solids are called the tetrahedron, the cube, the octahedron, the dodecahedron, and the icosahedron. One can easily see why the ancient Greeks, and in particular the Pythagoreans, were attached to them. Besides the sphere, they are the most symmetric of all solid figures.

Probably no one before Kepler had considered it much of a coincidence that there are six planets and five Platonic solids. It took both a mystic and a scientist to see a connection and, more importantly, to work out a plausible hypothesis for why it was so. For Kepler realized that if there are six planets, there are five *spaces between planets*. If one imagines the planets living, in Aristotelian fashion, on crystalline spheres, there is just enough space between Mercury and Venus to wedge in an octahedron, just touching Mercury's sphere on the inside and the sphere of Venus on the outside. Similarly, there is

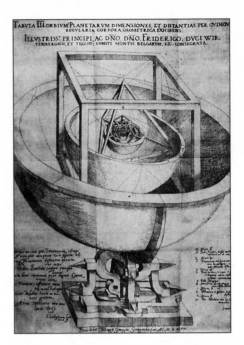

Kepler's solar system model from his *Cosmographic Mystery*, published in 1595. Kepler explained the sizes of the planetary orbits by a system of nested solids, with differently shaped solids separating the orbital shells of the planets. This scheme could be reconciled with astronomical observations only if the Sun was at the center—in agreement with Copernicus's controversial theory. *Photograph courtesy of Linda Hall Library of Science, Engineering, and Technology.*

just enough room between Earth and Venus to squeeze in an icosahedron; enough between Earth and Mars to fit a dodecahedron; enough between Mars and Jupiter for a tetrahedron; and enough between Jupiter and Saturn for a cube.

Of course, Kepler couldn't figure this all out at once. He spent the next year working out the details, and it cannot have been easy. The fit was not perfect; he had to fudge his model a bit by making the crystalline spheres rather thick. But on the whole, Kepler's theory "predicts" the radii of all the planetary orbits correctly to within 5 percent, and this certainly seemed close enough to him at the time. He gave his first book a grandiose title that reflects his enthusiasm: *The Cosmographic Mystery of the Admirable Proportions of the Celestial Orbs.*

To understand Kepler's lifelong attachment to this theory, you have to realize two things: it is utterly Pythagorean, and utterly Copernican. The polyhedral theory is Pythagorean in the sense that it posits a numerical coincidence as the fundamental secret of the universe. It is Copernican because it requires the planets to occupy concentric orbits, with the Sun at the center. It is totally incompatible with Aristotle, who placed seven (not six) crystalline spheres around Earth. Moreover, in the Aristotelian model the radii of the orbits would be all wrong. Only when the distances are measured from the Sun, as in Copernicus's theory, do the polyhedra nestle together snugly.

It is only fair to say at this point that modern scientists consider Kepler's polyhedral theory to be nonsense—naive at best, embarrassing at worst. It is a perfect example of pure mathematics run amok, an elegant theoretical edifice that has no physical basis, no observational evidence (who has ever seen a dodecahedron floating in space?), and no causative mechanism. What's more, we now know of three more planets—Uranus, Neptune, and Pluto. Without the magical coincidence of six planets and five solids, the whole premise of Kepler's system collapses.

Because Kepler's polyhedral theory turned out to be wrong, science historians often gloss over it as a youthful error. But Kepler himself remained proud of it to the end of his life, and in fact the events of 1610 even gave him a chance to embellish it. We should not pretend the theory never existed simply because it didn't work out; it was fundamental to Kepler's thought.

For Kepler's contemporaries, *The Cosmographic Mystery*, published in 1595, was an impressive piece of work. The faculty senate at Tübingen approved it unanimously for publication, with one proviso: he had to remove a chapter in which he explained how Copernicus could be reconciled with the Bible. The prorector of the university gave Kepler a warning that was almost identical to one that Galileo would receive from the Inquisition twenty years later: it was okay to treat the Copernican doctrine as a mathematical hypothesis, a working model that might give accurate results without being literally true. But it was definitely inappropriate as a guide to interpreting Scripture.

The Cosmographic Mystery brought Kepler into contact with two other scientists who would have a profound impact on his career: the astronomer Tycho Brahe and the mathematician Galileo Galilei. Brahe, who was at that time the Imperial Mathematician for Rudolph II, invited Kepler to work with him, an invitation that Kepler eventually accepted in 1600. When Brahe died the following year, it was Kepler who inherited his title. More importantly, Kepler also inherited twenty years of astronomical data that Brahe had compiled at his observatory on the island of Hven in Norway, the most accurate observations of the pretelescopic era. For someone

Galileo Galilei (1564–1642). His arguments on behalf of the Copernican solar system, in which Earth revolves around the Sun, led him into conflict with the teachings of the Catholic Church. In a famous trial in 1633, he was forced to recant and placed under house arrest for the rest of his life. Ironically, Kepler had chided Galileo as early as 1597 for failing to commit himself publicly to Copernicanism. It was only Galileo's discovery of the moons of Jupiter in 1610 that persuaded him to enter the fray. *Photograph courtesy of Library of Congress.*

who had difficulty observing the heavens himself, it was a godsend. On the basis of Brahe's data—and in particular his observations of Mars—Kepler would ultimately work out his three laws of planetary motion, which, unlike the polyhedral theory, are still taught in physics classes today.

The contact between Kepler and Galileo was much briefer and, to a lover of science, represents a great opportunity missed. In 1597, Galileo wrote to Kepler to compliment him on the book and to say that he, too, agreed with Copernicus's theory but that he had never dared to publish his own opinion. Kepler dashed off a heartfelt response full both of praise and reproach: "I would have wished, however, that you, possessed of such an excellent mind, took up a different position," Kepler wrote. "With your clever secretive manner you underline, by your example, the warning that one should retreat before the ignorance of the world. . . . You could help your comrades, who labor under such iniquitous criticism, by giving them the comfort of your agreement and the protection of your authority. . . . Have faith, Galilei, and come forward!" How ironic to see Galileo, whom history would remember as a martyr to scientific truth, chastised for "retreat[ing] before the ignorance of the world"! In any event, Galileo did not choose to come out of the closet, nor did he even reply to Kepler's letter. The two had no more contact until the fateful spring of 1610, when the roles were reversed, and Galileo would beseech Kepler for the "comfort of [his] agreement and the protection of [his] authority."

Projectiles and Planets

Before 1609, Galileo Galilei was even less of an astronomer than Kepler, and perhaps his reluctance to commit himself publicly to Copernicus's theory reflects a sensible caution in a field where he was not expert. He *was* the world's leading expert in the field of physics called mechanics, which deals with the way things move. While Kepler was dismantling Aristotle's cosmology, Galileo was systematically chipping away at Aristotelian mechanics. Galileo is best known now for an experiment that may or may not ever have taken place—dropping two balls of different weight off the Leaning Tower of Pisa, to show that they landed at the same time (thereby refuting one of Aristotle's theories). Most likely this was an embroidered version of the truth, which is that Galileo spent many years carefully measuring how things move—sliding blocks down inclined planes, swinging pendulums, and so on.

Some of Galileo's ideas defied common sense. For example, Aristotle had believed that moving bodies always come to a stop unless there is some force pushing on them. Who could disagree with that? If you're driving a car and shift into neutral, the car will eventually coast to a stop. No matter how hard you hit a golf ball, it eventually comes to rest—even if you're Tiger Woods. But Galileo realized that none of our everyday experiences truly involve objects with *no* force on them. There is always a force of friction. The less friction there is (think of a car slipping on an icy road), the longer it takes to stop. With great daring, he concluded that if there were *no* friction *at all*, the car would *never* stop. A moving body with no force on it keeps moving forever—exactly the opposite of what Aristotle said.

Having deduced how objects move when there is no force on them, Galileo also discovered how they move when there is *one* constant force on them, namely the force of gravity. Then the path is a graceful arc called a parabola. This is more or less the trajectory of a golf ball in full flight. But again, don't take this assertion too literally. The path of any projectile on Earth is distorted, at least a little bit and sometimes a lot, by friction with the atmosphere. The parabola is only an ideal, which would be achieved if the atmosphere offered no resistance—a golf ball on the Moon, for example.

Meanwhile, as Galileo was working on the motions of all sorts of projectiles, Kepler was trying to understand the orbit of the planet Mars. His "war with Mars," as he called it, lasted from 1601 to 1606. It began when Tycho Brahe asked Kepler to work out the motion of the red planet, which his assistant Longomontanus had been unable to do. Kepler boasted that he would be able to do it in eight days.

At first Kepler tried to fit its orbit to a circle about the Sun. Even the Aristotelians had realized that the Sun was not at the exact center of the planets' orbits. So Kepler was free to try circles with different centers. No matter how he tried, though, he couldn't find a circle that fit Brahe's observations of Mars. He came close; the best circle he found misrepresented Mars's position by no more than eight minutes of arc. For comparison, this is roughly the size of a quarter seen from thirty-six feet away. You or I would be very hard put to tell, with our naked eyes, that Mars was out of its predicted position by such a small distance. But Kepler did not believe that Brahe could have made a mistake this large.

What happened next was a decisive moment in science history. Kepler did what Longomontanus would never dream of doing: He threw out many centuries of tradition, on the basis of one tiny discrepancy with an experiment. He practiced as Galileo would one day preach, letting nature rather than textbooks have the final say. Since a circle would not work, the orbit of Mars must be some sort of oval. And over the course of five years, with many false starts and mistakes, he worked out exactly what kind of oval it was: an ellipse. This is the same shape as a circle viewed from an oblique angle. Moreover, Kepler worked out that the Sun was not at the center of the ellipse but at a focus, one of two special points used in the mathematical definition of an ellipse. (In a circle, the distance from any point to the center is a constant. In an ellipse, the *sum* of the distances from any point to the two foci is a constant.)

To readers who have never fallen under the spell of mathematics, finding the exact shape of a golf ball's trajectory or a satellite's orbit may not seem like a very big deal. What is so special about Galileo's parabola, or Kepler's ellipse? Why wouldn't any arch-shaped figure do for the trajectory of a ball, and any oval-shaped figure do for the trajectory of a planet? Particularly since in the real world there *are* always small distortions—the atmosphere distorts the trajectory of a ball, and the gravity of other planets distorts the trajectory of Mars.

First, to a Galileo or a Kepler or even to a mathematician today, the particular shapes *do* matter. Parabolas and ellipses had been studied by the ancient Greek geometers and were part of the education of any scientist. To encounter them again here was like seeing an old friend in a new town. If the planets couldn't move in circles, it must have been reassuring that they at least moved in a curve that had been known since antiquity. More than reassuring, it must have seemed fitting for a well-planned universe. Galileo once expressed the very Pythagorean belief that the book of nature "cannot be read until we have learned the language and become familiar with the characters in which it is written. It is written in mathematical language, and the letters are triangles, circles, and other geometrical figures, without which means it is humanly impossible to comprehend a single word."

What's more, the ellipse and the parabola could have given Kepler and Galileo a clue to a fact that neither man suspected: *they were both working on the same problem.* A parabola is just a special form of an ellipse—mathematically, an ellipse stretched out to infinity. The only reason the golf ball appears to travel in a parabola instead of an ellipse, like Mars, is that the golf ball is much closer to the body it is "orbiting." If the ground didn't get in the way, the golf ball's orbit would be revealed as a highly elongated ellipse. Because Kepler and Galileo didn't catch on to this, it would be left to Isaac Newton, at the end of the seventeenth century, to make the connection between projectiles on Earth and projectiles in space.

More pragmatically, what both Galileo's and Kepler's work allowed for the first time was precise numerical prediction. If all you know is that a ball moves in some sort of arc, you can't tell *exactly* where it will land. If you know only that a satellite moves in some sort of oval, you can't tell *exactly* where to point your GPS receiver, and you certainly have no business sending men to the Moon. Kepler and Galileo began a quantitative revolution in science. Kepler's refusal to accept a measly eight minutes of error was, perhaps, the first shot of the revolution.

1610: Act One

After years of traveling in separate orbits, Kepler and Galileo began to converge again in 1609. In that year, Kepler finally saw into print his book *New Astronomy*, in which he announced his first two laws of planetary motion. (The first is the elliptical orbit; the second is a precise version of the statement that planets speed up as they get closer to the Sun and slow down as they move away.) Casting about for something to do next, he went back to his "lunar thesis" of 1593, polished it up, added some new facts he had learned about planetary movements from his battle with Mars, and recast it into a new narrative form that he hoped would make it more reader-friendly. He undoubtedly would have published it within two or three years, had some other strange events not occurred.

The first was the invention of the telescope. The first working models appeared in Flanders in 1608. Over the next two years, they spread like wildfire through the intellectual circles of Europe—the Rubik's Cube of their era, only much more useful. Galileo learned the secret in 1609, and by the end of the year he had made the best spyglass in Europe; it could make objects seem twenty times closer. At first he was content to demonstrate to his employers its value on Earth. He showed the doge of Venice how it could be used to spy incoming ships at sea long before they could be seen by the naked eye. His reward for this was quick and gratifying: the Venetian senate voted to double his salary.

But in November 1609, Galileo began to point his telescope at a more lofty target, the Moon. He was not the first scientist to look at the Moon through a telescope; in England, Thomas Harriot had done so as early as August. Harriot's pupil Sir William Lowe had written that it looked "like a tarte that my cooke made me the last weeke—here a vaine of bright stuff, and there of darke, and so confusedlie all over."

But where others saw only a confusing pattern of light and dark spots, Galileo understood that he was seeing a *landscape*. The Moon's detail showed up most clearly in the boundary region between the illuminated part of the Moon and the dark part. This dividing line is called the terminator, and it is the place where, to a Moon-dweller, the Sun would be seen setting or rising at any given moment. Sunrise and sunset are the best times to photograph a landscape on Earth, because they are the times of greatest shadow and contrast, and the same is true on the Moon.

Seen through the telescope, the terminator visibly moves during the course of an evening, and the changes Galileo saw in the patterns of light and dark confirmed his belief that the Moon had a rugged

surface. Some bright spots would light up even before the terminator reached them; these spots would be illuminated on the side facing the Sun, and a dark shadow on the other side would persist for a
while after the terminator moved by. Of course, this is exactly what
happens to a mountain peak at sunrise: the side facing the Sun
lights up, and the other side stays in shadow.

There were other features, in great profusion, that had the opposite behavior: the side opposite the oncoming Sun would light up
first, and the nearer side would light up last. These features Galileo
took to be valleys. He was a little puzzled by their circular appearance, so different from most valleys on Earth, but he compared one
of them to Bohemia, a flat land surrounded by hills. Of course, they
were not valleys but craters. Nowadays, in our posttelescopic, postas-
tronautic era, it is common knowledge that the Moon is pocked with
craters, but Galileo was seeing them for the first time.

For the most part, Galileo's first descriptions of the Moon show
admirable restraint. He resisted the temptation to give in to conjecture, and instead described exactly what he saw—not a mountain,
but a shadow that behaved in such-and-such a way, until the case
was clear and compelling that the shadow had to be a mountain.
Galileo noticed, for example, that the dark-colored regions of the
Moon seemed to be lower than the lighter regions, based on his
observations of the terminator. He hints at a possible interpretation
but falls far short of endorsing it: "If anyone wanted to resuscitate
the old opinion of the Pythagoreans that the Moon is, as it were,
another Earth, its brightest part would represent the land surface
while its darker part would more appropriately represent the water
surface." We will see that Kepler, by contrast, was eager to "resuscitate the old opinion." For a few decades, at least, most scientists
agreed that the dark regions were actual seas of water, and this is
why they gave them such names as Mare Tranquillitatis (Sea of Tranquility, where *Apollo 11* landed).

In our casual classification of scientists into "loonies" and "non-
loonies," Galileo definitely counts as a nonloony. As 1610 began, his
attention moved elsewhere, and he never again paid very much attention to the Moon. But even in the two months or so he spent on it,
he gave a masterful performance. He resolved an ancient debate. Yes,
the Moon is a solid world—not a fire, not a puncture in a cosmic
inner tube, not a god, not a mirror made of "quintessence." He also

These engravings from Galileo's *Sidereus Nuncius (Starry Messenger)*, published in 1610, were the first published observations of the Moon through a telescope. They provided an incontrovertible proof of Anaxagoras's theory that the Moon is a world like our own, with mountains and valleys. A crucial feature of Galileo's pictures was his careful attention to light and shadow on the Moon's surface; note the superb portrayal of craters (exaggerated in size). *Photograph courtesy of Linda Hall Library of Science, Engineering, and Technology.*

pioneered a scientific, experimental method for comprehending the Moon's surface by studying the terminator. He discovered a previously unknown phenomenon, the Moon's craters. He drew gorgeous ink-wash pictures of the Moon that would be the best available for another forty years.

And then, on January 7, 1610, Galileo noticed something else through his telescope. Near the planet Jupiter, he saw three little stars all in a straight line. A curious constellation, worthy of a passing mention in his journal. But then, on the following nights, something astounding happened: the stars *moved*. They rearranged themselves into different positions. A fourth one appeared. And they moved *with* Jupiter, as if dancing attendance on the king of the gods. Over time he realized that he was seeing four new planets or moons that were circling around Jupiter, but appeared to be in line with it because he was seeing them from the side. The inner one moved fastest, and the outer one moved the slowest. It was like a solar system in miniature—a *Copernican* solar system.

This was the revelation that finally persuaded Galileo to make his views public. He wrote, "We have an excellent and splendid argument for taking away the scruples of those who, while tolerating with equanimity the revolution of the planets around the Sun in the Copernican system, are so disturbed by the attendance of one Moon around the Earth while the two together complete the annual orb around the Sun that they conclude that this constitution of the universe must be overthrown as impossible." Perhaps this same doubt had even crossed Galileo's mind? But now the Moon was no longer an oddball; it was one of many moons.

At this point Galileo knew he had to act fast. Other scientists would surely procure or make telescopes of their own, and some were sure to train them on Jupiter. The discovery of the ages would be his only if he published his observations as quickly as possible. On March 13 (two days before the rumor reached Kepler), Galileo's first book on astronomy, containing his observations of the Moon and the moons of Jupiter, started rolling off the presses. *Sidereus Nuncius,* which translates as *Starry Messenger* or *Starry Message,* was nothing more than a pamphlet, filling only forty pages in Galileo's collected works. But it made him into an international celebrity, the Albert Einstein or Stephen Hawking of his time.

Kepler received his copy of *Sidereus Nuncius* on April 8, delivered personally by the Italian ambassador to the Holy Roman Empire, Giuliano de' Medici. We can imagine his joy at reading it. He must have relished Galileo's unabashedly pro-Copernican stance and exclaimed with delight when he found out that the new "planets" were not orbiting a distant star. He had not lost his argument with Wackher after all. He was also spared an even worse embarrassment: if the new planets had been orbiting the *Sun,* then he would have had to rethink his whole polyhedral theory. All in all, the fact that the planets were orbiting Jupiter was the best possible news.

There was better news to come. On April 13, he had a personal audience with de' Medici, and the ambassador told him that Galileo wanted Kepler's response to the pamphlet in writing. But the next courier would be leaving for Italy in only a week. Could Kepler write it by then?

The reason Galileo needed Kepler's imprimatur was simple: he anticipated a backlash, and he did not have long to wait. Near the end of April he experienced a fiasco when he attempted to demonstrate the moons to an astronomer named Giovanni Magini at Bologna. His telescope had a very narrow field of view, and the inexperienced people who looked through it professed not to see anything at all. Other scholars would soon say that the so-called moons of Jupiter were optical illusions, and some would refuse point-blank to look in the telescope—an amazing attitude of "If I can't see it, it isn't there." For a few months, at least, Galileo would have been quite alone if it hadn't been for Kepler.

Kepler penned his reply in six days and brought it back to de' Medici on April 19, unsealed. He must have worked almost nonstop, for his response is almost as long as the *Starry Messenger* itself. The following month he published his open letter to Galileo as *Conversation with the Starry Messenger.*

Kepler starts out by making one thing clear: he hasn't looked through a telescope yet himself. (He wouldn't get a chance until August.) Nonetheless, he is prepared to vouch for the authenticity of Galileo's discoveries. "But why should I not believe a most learned mathematician, whose very style attests the soundness of his judgments? . . . Shall I with my poor vision disparage him with his keen sight? . . . Shall I not have confidence in him, when he invites everyone to see the same sights, and what is of supreme importance, even offers his own instrument in order to gain support on the strength of those observations?"

After this stirring beginning, Kepler squarely addresses the question of whether the images seen in a telescope are real. Not only does he understand how the telescope works, he practically gives a blueprint for a better one. This part of Galileo's *Messenger* has clearly not surprised him one whit. The Keplerian telescope, with two convex lenses, did in fact replace the concave-convex style that Galileo used.

From there, Kepler launches into the heart of the matter, the astronomy. Where Galileo wrote with cool, clinical detachment, Kepler burns with passion and enthusiastic speculation. Reading the *Starry Messenger* and the *Conversation* together is like listening to two very different musicians, one introducing a theme and the other developing it with wild improvisations.

First, since the telescope *does* work and doesn't show a highly magnified fog, Kepler writes that "we must virtually concede, it seems, that that whole immense space [between Earth and the Moon] is a vacuum," and with this yet another pillar of Aristotelian physics crashes down, for Aristotle had written that "nature abhors a vacuum."

Further, Kepler predicts that Mars will have two moons—correct again, but for utterly wrong reasons. It's Pythagoreanism again. If Jupiter has four moons and Earth one, then Mars must have two and Saturn eight to make a nice geometric sequence: 1, 2, 4, 8. He even invents a clever new polyhedral theory to account for Jupiter's moons: The three spaces correspond to the three rhombic solids—the cube with six faces, a cuboctahedron with twelve, and an icosidodecahedron with thirty. (The latter two are regular polyhedron "wanna-bes": their faces are all identical, their sides all have the same length, but the faces are diamonds, not squares.) He maintains that Galileo's discoveries do not upset Earth's special place in the universe: "We humans live on the globe which by right belongs to the primary rational creature, the noblest of all the (corporeal) creatures." Jupiter's moons, he says, are a consolation prize for its inhabitants, who are too far from the Sun to see Mercury and perhaps Venus. We humans are the only ones who can see the polyhedral solar system in all its glory.

Finally, Kepler makes a prediction that must have seemed like utter fiction: "As soon as somebody demonstrates the art of flying, settlers from our species of man will not be lacking. . . . Given ships or sails adapted to the breezes of heaven, there will be those who will not shrink from even that vast expanse. Therefore, for the sake of those who . . . will presently be on hand to attempt this voyage, let us establish the astronomy, Galileo, you of Jupiter, and me of the Moon."

1610: Act Two

The above passage alone, I think, qualifies Kepler as an honorary loony, even if his true passion was always for the Copernican system. In fact, he had been thinking a great deal about the Moon, and was almost ready to print his own lunar astronomy. He even refers to it once or twice in the *Conversation*. But events continued to conspire against him, and the book, called *Somnium (The Dream)*, actually did not appear in print until 1634. However, it seems appropriate to examine it now, because it was composed in the shadow of the events of 1609 and 1610.

Somnium has been called the first work of science fiction, but most of it is serious science, though conjectural at times. As Kepler says in a footnote written later: "The objective of my *Dream* was to work out, through the example of the Moon, an argument for the motion of Earth." Even today, it is full of fascinating and original insights.

Kepler's lunar inhabitants call the Earth Volva because it revolves. (Remember that the whole book grew out of his desire to prove this point.) "During the night, which lasts as long as fourteen of our days and nights," Kepler writes, "[Volva] appears not to move at all from its place, unlike our Moon, it nevertheless turns like a wheel in its place." Elsewhere, he writes, "Volva stands in the sky as if fixed by a nail." This is basically correct: the only reason why the *Apollo 8* astronauts could take their celebrated photographs of earthrise is that *they* were traveling around the Moon.

Kepler also writes that seasons on the Moon are less dramatic than on Earth, because the tilt of the Moon's axis is less (5 degrees as opposed to 23 degrees). He would have been right, except for the fact that the Moon has no atmosphere and therefore no seasons.

Further, Kepler notes that the equinox on the Moon doesn't always occur at the same time of year. Again, this is a subtle deduction based on impeccable astronomical reasoning. Kepler knew that the Moon's orbit is tilted with respect to the ecliptic plane. The points where it intersects that plane are called "nodes." The Moon's equinoxes happen when the line of nodes points directly toward the Sun. (This is also when lunar and solar eclipses are seen on Earth.) But the line of nodes also turns at a rate of one rotation every 18.6 years, and this causes lunar equinoxes to migrate as well—a point that would surely be important to farmers on the Moon, if there were any.

Speaking of eclipses, Kepler beautifully describes what they would look like from the Moon. "Eclipses of the Sun are common; they do not ever see a total eclipse of Volva, but they see a certain small spot, which is ruddy on the edges and black in the middle, go across the body of Volva; this spot enters from the east of Volva and leaves by the western edge, taking the same course in fact as the spots [i.e., continents] that are native to Volva, but surpassing them in speed."

When Kepler errs, it is only in trying too hard to make the Moon habitable. His description of the surface inspired many later science fiction writers, such as H.G. Wells, but turned out to be mistaken: "The whole of it . . . is porous and pierced through, as it were, with

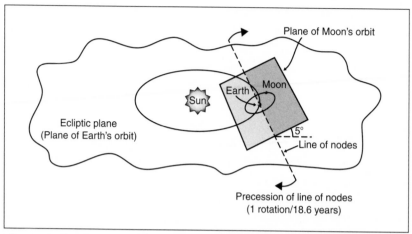

The precession of the nodes was a long-known astronomical phenomenon that got a new interpretation in Kepler's *Dream*. Naively, one might expect solar eclipses to happen every new Moon, because that is when the Moon is between Earth and the Sun. But in reality, they happen only about twice a year, when the *line of nodes* (the intersection of the Moon's orbital plane and the ecliptic plane) points toward the Sun. At other times of the year, the Moon is either south or north of the ecliptic at new Moon, so no eclipse occurs. Kepler realized that the alignment of the nodes with the Sun would represent an *equinox* for someone living on the Moon. On Earth, the fall and spring equinoxes always happen at the same time of year, but on the Moon they would gradually vary because the line of nodes does not remain fixed; it rotates once every 18.6 years.

hollows and continuous caves." The craters, he thinks, are reservoirs built by the inhabitants: "The perfectly round shape of the cavities and the arrangement of them ... is something artificial and the result of some architectural intellect." Kepler would not be the last scientist to be fooled by the Moon's craters.

Prophetically, Kepler writes: "The most pleasant thing of all [in the Moon] is contemplation of Volva, the sight of which the dwellers there enjoy in place of our Moon." Indeed, when the Apollo astronauts went to the Moon, many of them commented that the beautiful blue and green Earth was the most impressive spectacle they saw.

1610: The Epilogue

With these three books—Galileo's *Starry Messenger,* and Kepler's *Conversation with the Starry Messenger* and *The Dream*—humanity's knowledge of its nearest neighbor took a quantum leap forward. The story remains to be told of how humanity reacted to that knowledge.

During the summer of 1611 Kepler started circulating his Moon manuscript to his friends, and somehow a bootleg copy got out. One of the places the pirated version made its way to was Leonberg, where his mother, Katarina, was now living.

Earlier, I mentioned briefly that Kepler had written a fictional prologue to *The Dream* in order to make it more appealing to a broad audience. Unfortunately, as a first-time fiction writer, he hadn't bothered to change it very much from the story of his own life. The "dream" of the title is about a man from Iceland named Duracotus. The protagonist had never known his father, and had been raised by his mother, Fiolxhilde, an herbalist. He comes back home after studying with the great Danish astronomer Tycho Brahe. (Is this starting to sound familiar?) His mother tells him that she can summon a "spirit" who can actually transport him to the Moon. She invokes the "Daemon from Levania" (a made-up name based on the Hebrew word for "moon") with a rune of twenty-one letters. The rest of the book is basically a tour guide to the Moon with the Daemon as narrator, including the various points I have just listed.

Now consider how this story played in rustic Leonberg. This was the early 1600s, and witch hysteria was running rampant. In 1615, six "witches" were executed in Leonberg alone, a town of a few hundred inhabitants. Katarina was by then a cantankerous old woman, and an herbalist. Kepler's attempt at fiction was too transparent for his own good; the townspeople easily recognized him as Duracotus, and Katarina as Fiolxhilde. They couldn't care less about the scientific message in the book. To them, Kepler's book, with the summoning of the Daemon, looked just like a confession of witchcraft. An especially damning passage was the one where the Daemon says what kinds of people are able to make the journey to the Moon: "Especially suited are dried-up old crones, who since childhood have ridden over great stretches of the earth at night in tattered cloaks on goats or pitchforks." Much later, in an annotation to the book, Kepler would lament, "Such a desire had I to jest and to argue jestingly!"

But it was no laughing matter in a country that was on pins and needles. Throughout 1615 the accusations grew against Kepler's mother, and in 1616 a warrant was issued for her arrest, which eventually grew to include forty-nine separate charges. It hung over her head for four years as the local magistrates gathered evidence and

Kepler fought to clear her name. To no avail: in 1620, Katarina was arrested and spirited off to jail in a closed chest, to keep anyone from hearing her cries for help. When Kepler next saw her, she was chained up in a prison cell.

But Johannes, who had once counseled Galileo not to retreat before the forces of ignorance, wasn't about to do so now. He left his post in Linz, Austria (where he had moved after Emperor Rudolph had died), telling no one his whereabouts. He called on his old friend Christoph Besold, the same Besold with whom he had planned the "lunar thesis" years earlier, to serve as his mother's advocate. He drew up point-by-point rebuttals to all the charges against his mother. The reaction of the magistrates who had expected a straightforward trial is eloquently expressed in the court records: "The prisoner appears, alas, with the support of her gentleman son, Johannes Kepler, the mathematician."

In the end, Kepler created enough of a furor to keep his mother alive. The case was referred to the law faculty in Tübingen. They did not exactly acquit Katarina, but they decided that she was to be taken to the torture chamber and threatened with the rack. If she did not confess, she was to be released. Of course, the tough old lady, the model for Fiolxhilde, did not crack. In October 1620, at age seventy-four, she became a free woman again. The following year she died.

How much Kepler had suffered for his innocent lunar thesis! By then, his country was suffering, too. The Thirty Years' War, a war of religious intolerance between Protestants and Catholics, had begun in 1618. Kepler would not live to see its end. In the final years of his life he was a man without a permanent home, trying desperately to take care of unfinished business. For starters, there was a table of Tycho Brahe's astronomical data that he had promised the emperor to edit. The emperor was long dead, but a promise was a promise. In 1629 the *Rudolphine Tables* were finally published.

And then there was one personal piece of unfinished business: the lunar geography. Kepler annotated the ill-starred manuscript copiously this time, making sure that no one could miss all of its meanings, hidden or otherwise. Duracotus, he said, represented Science. The mother, Fiolxhilde, represented Ignorance. The absentee father was Reason, for Science is the child of Reason and Ignorance, and "it is natural that this father should be either unknown to the mother or concealed to her." The rune that Fiolxhilde used to invoke the Daemon of Levania? It was "Astronomia Copernicana."

So that is how Copernican astronomy led Kepler to the Moon. He never did get to see his lunar thesis in print: he died in 1630, and his son Ludwig finished the job for him in 1634.

Astute readers will recall that Galileo was undergoing a trial of his own at about the same time. In 1633 he was summoned to Rome and placed on trial by the Inquisition, for advocating Copernicanism in his book *Dialogue concerning the Two Chief World Systems*. The story of this proceeding, of Galileo's "confession" and forced recantation, has been told many times and from many points of view. Earlier historians tended to interpret the trial as an inevitable collision between "ignorant" religion and "enlightened" science, but modern scholarship emphatically denies this simplistic view. Galileo the scientist was a deeply religious Catholic, and mortified by the fact that the church had picked a fight it could only lose. The Holy Office was perfectly aware of Galileo's immense reputation, and treated Italy's most famous scientist with velvet gloves (by Inquisition standards).

Galileo's trial, like those of Anaxagoras and Katarina Kepler, was influenced by a host of political factors other than the conflict between science and religion. Humans have always been very creative at dividing themselves into Us and Them. Kepler and Galileo were unlucky enough to live in a place and at a time when the quarrel between Us and Them (in this case, Catholics and Protestants or vice versa) was especially intense. Perhaps in another era, the Copernican revolution could have taken place less painfully, as indeed it almost did in ancient Greece.

More Popularizers

In spite of the trouble they had caused, the books of Galileo and Kepler found an enthusiastic audience and inspired a number of copycats. The seventeenth century saw many stories of fantastic voyages to the Moon, as well as scientific or semiscientific books that discussed the questions of whether the Moon was inhabited and whether men could fly there. Though Galileo's discovery of the moons of Jupiter may have been more momentous scientifically, the world of our Moon gripped the public imagination in a way that Io, Europa, Callisto, and Ganymede did not.

The first literary mentions of the new discoveries came as early as 1611, in a novel by John Donne and a play by Ben Jonson. A decade later, Jonson wrote a play called *News from the New World Discovered in the Moon*. However, the tempo of literary Moon voyages really picked up after Kepler's *Dream* was published. In 1638 Francis Godwin published *The Man in the Moone: or a Discourse of a Voyage Thither,* and the scientist John Wilkins published a scientific treatise with the lengthy title *A Discourse Concerning A New World and Another Planet: The First Book, The Discovery of a New World; or, A Discourse tending to prove, that 'tis probable there may be another habitable World in the Moone.* Already, Wilkins enumerates four ways by which one might travel to the Moon: "1. By spirits, or angels." (Remember Kepler's Daemon!) "2. By the help of fowls. 3. By wings fastened immediately to the body. 4. By a flying chariot." In France, Cyrano de Bergerac picked up the theme in 1656 with his *Comic History of the States and Empires of the Moon.* Like many of the cosmic voyages, it starts with the narrator trying to convince his skeptical friends that "the Moon is a World like ours, to which this of ours serves likewise for a Moon." To prove his point, he builds a flying machine. His first attempt is unsuccessful and gets him only as far as Canada, but then some soldiers, for their amusement, attach some fireworks to the machine, providing it with the extra oomph it needs to reach the Moon. Once there, Cyrano faces some unexpected problems: he is put on trial for claiming that *Earth* is inhabited! Thanks to de Bergerac, the Moon once again becomes a laughing matter.

Astronomers, too, were busy with the new world that Galileo and Kepler had revealed, drawing maps and naming the new features they discovered. In this way, too, the Moon nearly became a battleground between Protestants and Catholics, as each side chose names after its own heroes. However, in 1651 it finally dawned on a Jesuit priest named Giovanni Battista Riccioli that the Moon belongs to all people, and he devised the nonsectarian naming scheme we use today. The Moon's features are named in Latin, the closest thing to a common language among literate people in the Western world. The seas are named after effects that had been attributed to the Moon in myth or legend: thus Mare Fecunditatis (Sea of Fecundity) or Mare Crisium

(Sea of Madness). The craters are named after scientists and scholars, including many of our old favorites—Anaxagoras, Plato, Aristotle. With remarkable magnanimity, given the fact that his church had banned the Copernican theory, Riccioli named two of the brightest craters after Aristarchus and Copernicus. But just in case anyone suspected him of being a Copernican sympathizer, he also banished Kepler and Galileo to small craters in the Ocean of Storms. Generosity, it seems, still had its limits.

4

The Clockwork Solar System

"The sky is falling! The sky is falling!" cried Chicken Little in the children's fairy tale. She wasn't so much wrong as ahead of her time. Everything in the sky *is* falling—but it took humans until 1686 to become aware of it.

That was the year that Isaac Newton published his masterpiece *Mathematical Principles of Natural Philosophy* (usually known by its Latin title, *Principia Mathematica*) and explained to the world what "falling" really means. It means moving under the influence of a gravitational field. That's what a ball does when it falls, or an acorn like the one that hit Chicken Little, or an apple like the one that supposedly (if we are to believe scientific fairy tales) inspired Newton. The Moon is falling, too. In fact, the Moon's motion is perhaps a purer example of falling than any of the others. The ball, the acorn, and the apple are all impeded by the atmosphere, while the Moon is not.

If this seems a little bit far-fetched, keep in mind that there is a big difference between *falling* and *landing*. The Moon is always in a state of free fall, always bending its course toward Earth, but it will never land. Its momentum keeps carrying it far beyond any landing place on Earth. Though the distinction between falling and landing seems simple, one might argue that this is precisely what confounded Aristotle, Kepler, and centuries of physicists between them.

Newton comprehended the difference, not only in words but also in precise formulas that have made the motions of the planets predictable and space travel possible. Before Newton, the solar system was like a watch that astronomers could not pry open; they could only watch its hands go around. After Newton, they could open it up and study the gears and springs. From then it was only a short time before they started to ask how it was made.

Newton's Theory of Gravitation

In the first part of the seventeenth century, Johannes Kepler had taken the astronomy of his idol Nicolaus Copernicus beyond anything Copernicus ever imagined. Kepler replaced Copernicus's circles with ellipses, established the Sun at the focus (not the center) of the ellipse, and brought an unprecedented precision to the analysis of planetary orbits. He also anticipated Newton by trying to make the Sun into the cause of planetary motions, a goal reflected in the title of his 1609 book *A New Astronomy Based on Causes.*

Unfortunately, the cause Kepler proposed did not conform very well to reality. He thought that the Sun continually propelled the planets onward in their orbits by some kind of magnetic force. Few scientists believed this, but many subscribed to another theory based on the same misconception. In 1644, the French mathematician and philosopher René Descartes published his "vortex theory," according to which the universe is divided into different rotating regions. The planets are swept along by the Sun's vortex, while a smaller eddy within that vortex carries the Moon around Earth. Both Descartes and Kepler were under the impression that a planet moving in a circle (or an ellipse) needs a force to push it in the direction it is moving. That is wrong, by exactly ninety degrees. The planets do not need any force to propel them *forward,* but they do require a force to pull them *inward,* toward the Sun.

Before proceeding with the history, it's worth explaining this point, which is very far from obvious to anyone but a physicist. Imagine tying a stone to the end of a rope and swinging it about your head. You will notice that the rope points straight from the stone to the center of the circle it travels in. It shouldn't be too hard to convince yourself that a rope is incapable of exerting a force in any direction except *in the direction of the rope.* Thus the rope is pulling inward, not forward, on the stone. You don't have to take my word for it, or anybody else's. The rope doesn't lie.

However, if you do want somebody's word, there would be no one better to ask than Newton. In *Principia Mathematica* he wrote: "The change of motion is proportional to the impressed force and takes place in the direction of the straight line in which the force is im-

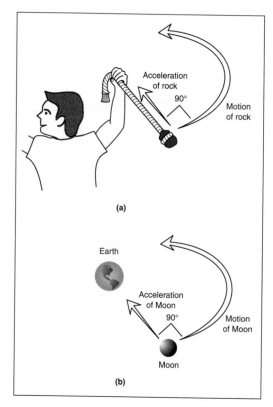

(a)

(b)

When a person whirls a stone in a circle, as in (a), the force he exerts on the stone is *perpendicular* to the direction of motion. Isaac Newton realized that the same reasoning could be applied to Earth and the Moon (b). Even though the Moon moves *around* Earth, the force on it points *toward* Earth. Thus, Newton reasoned, the force that keeps the Moon in orbit is exactly the same as the force that pulls an apple to the ground. In a nutshell (or an apple core), that is the law of universal gravitation.

pressed." The stone's direction of motion is always changing—it is turning toward the center of the circle. Newton says, then, that the force on the stone points *in the direction of the change of motion*—toward the center of the circle, *not in the direction of the motion itself.* This is the subtle point that Kepler and Descartes had missed.

The same reasoning applies to the Moon. There may not be a visible rope attached to the Moon, but the Moon's *change of motion* (or, to use the modern word, *acceleration*) is always toward Earth. It follows that the force Earth exerts on the Moon—the force that Newton called "gravity"—points directly toward Earth, too.

Just as Galileo has his experiment at the Leaning Tower of Pisa, Newton has his apple—a moment of insight that may or may not have occurred in reality but that has served for centuries as a sort of icon for the real breakthrough it represented. You don't have to be a genius to see that some kind of invisible force pulls an apple toward Earth. The real stroke of genius was to realize that there also is a force pulling the Moon toward Earth. Once he understood this, it was reasonable for Newton to suppose that the two forces were the same phenomenon. The Moon, just like the apple, is freely falling under Earth's gravity.

Going a step farther, Newton conjectured that *any* two masses in the universe—whether they are apples in his backyard in Woolsthorpe, England, or stars in the sky—attract each other in the same way. Larger objects create a stronger gravitational pull, which explains why we feel Earth's attraction but cannot feel our gravitational attraction to an apple at all. Nevertheless, the force is always there. In the twentieth century, physicists managed to measure the gravitational force even between ordinary, garden-size objects, thereby confirming Newton's law of gravitation to the best of their abilities.

Philosophically, Newton's "universal law of gravitation," which says that any two objects in the universe are attracted by the same kind of force, is very much in the tradition of Anaxagoras. Although Newton never claimed that heavenly objects literally come from Earth, his law does assert a deep kinship among all forms of matter. Every star, planet, and chunk of rock obeys the same physical laws. His law of gravitation forged the link between earthly objects and heavenly objects that Galileo and Kepler had missed. The *physical causes* of the parabolic motion of a golf ball and the elliptical motion of a planet, according to Newton's theory, were one and the same, even if the effects appeared to be different. In both cases, the physical cause was gravitation.

Beyond its qualitative difference from Kepler's and Descartes's theories, Newton's law also differed quantitatively. It said that the gravitational force falls off very rapidly with distance: If you move twice as far away from an object, your attraction to it decreases to one-fourth the original strength (1 divided by 2 squared). Three times as far away, the force of gravity diminishes to one-ninth (i.e., $1/3^2$) of its original strength. This rule is called the "inverse-square law." Kepler, on the other hand, had believed that the Sun's force on the planets waned in direct proportion to the distance.

The inverse-square law may seem like a minor technical point that only a math freak should care about. But if it weren't for the inverse-square law, you and I would be swept off the surface of Earth and sucked toward a fiery death in the Sun. Why? Standing on Earth's surface, you are about twenty thousand times closer to the center of Earth than to the center of the Sun. But the Sun is about three hundred thousand times more massive than Earth. So if gravity worked by simple proportions, your attraction to the Sun would be stronger than your attraction to Earth. But thanks to the inverse-square law, the Sun's force on you is less than one-thousandth ($300,000 \div 20,000^2$) as strong as Earth's. Such details make the difference between an inhabitable and an uninhabitable solar system.

Another important consequence of the inverse-square law, which was really not understood before Newton, is the possibility of escaping Earth's or the Sun's gravitational field. If gravity decreased in proportion to distance, it would take an infinite amount of energy to escape Earth's gravity. Since no rocket can carry an infinite amount of fuel, every rocket would have an absolute limit to how far away from Earth it could get before its fuel ran out. There would be a "glass ceiling" on our exploration of the universe. But because gravity decreases in proportion to distance *squared,* it becomes very weak at large distances—so weak that, with a large enough initial push, you can coast away as far as you want. Even now, the *Pioneer 1* and *2* spacecraft are coasting all the way out of the solar system, never to return. The old truism "What goes up must come down" is no longer true in a Newtonian universe. Instead, we should say that what goes up must either come down, or go up and up forever.

Amazingly, Newton deduced his inverse-square law without any experimental apparatus at all. How did he do it? The answer is that he knew Kepler's three laws of motion, which in effect turned our solar system into a gigantic scientific laboratory.

I have mentioned Kepler's three laws before, but now is the time to state them precisely:

1. Planets orbit around the Sun in ellipses, with the Sun at one focus. (Similarly, moons orbit planets in ellipses, with the planet at one focus.)

2. As a planet moves around the Sun, it "sweeps out" equal areas in equal times. For example, if you draw a triangle with vertices at the Sun, Earth's position on July 1, and Earth's position on July 2, it will have the same area as another triangle with vertices at the Sun, and Earth's positions on January 1 and January 2. This seems pretty obvious until you realize that Earth is *closer* to the Sun on January 1 than it is on July 1. Therefore, the January 1–2 triangle must be wider, to maintain an equal area. In other words, Earth speeds up in its orbit as it gets closer to the Sun.

3. The square of the amount of time it takes for a planet to make one revolution is proportional to the cube of the planet's distance from the Sun. For example, a year on Saturn lasts about 30 Earth years. The square of 30 is 900, and the cube root of 900 is about 9.6; hence we can conclude that Saturn is about 9.6 times farther from the Sun than Earth is.

Newton proved that Kepler's second law provides the *direction* for the gravitational force—it holds true if and only if the gravitational force is directed to the center of the gravitating body. This forever refuted Descartes's doctrine of a vortex sweeping the planets along. And Kepler's third law tells us about the *strength* of the gravitational force: Newton showed that this law is consistent with an inverse-square rule for the force, and no other. Together, the two laws fit together like a hand and a glove.

But there was still one test to pass. Did a centrally directed, inverse-square gravitational force predict that planets would orbit in ellipses? This problem defeated all the other leading scientists of England—Sir Christopher Wren, Edmond Halley (the discoverer of Halley's comet), and Robert Hooke (one of Newton's great competitors, particularly in working out the theory of optics). To solve it, Newton had to devise a whole new branch of mathematics, which we now call calculus.

What makes the problem so hard is that the force on a planet is constantly changing, both in direction and in magnitude, as it orbits the Sun. If you update the planet's position every day, you will still make mistakes in predicting the orbit, because the gravitational force on it will change during the day. Even if you update the orbit every hour, or every minute, you will still make mistakes, though they will get smaller and smaller. The only way to predict its orbit *exactly* was to develop a mathematics of continuous change. Newton called this the method of *fluxions,* an invented word that perhaps conveys the spirit of the subject better than "calculus."

With more than three hundred years of hindsight, it is nothing short of astounding that even after developing the rules of calculus, Newton arrived at a problem that was mathematically solvable. The odds were highly stacked against it. Mathematicians have since found

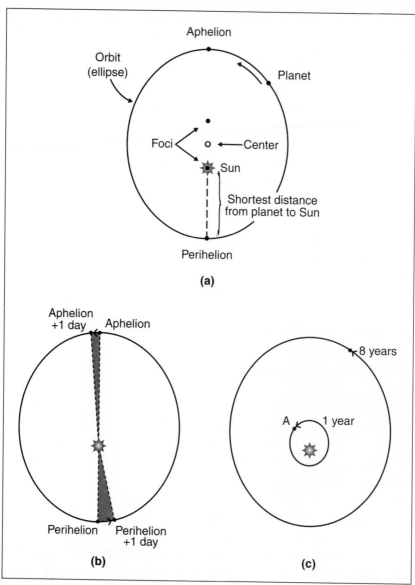

(a)

(b)   **(c)**

Kepler's three laws of motion—unlike his solar system model on page 36—have remained an accepted part of celestial dynamics. (a) A planet orbits the Sun in an ellipse, with the Sun at one focus. Note that there is no physical object either at the center of the orbit or at the other focus. (b) The speed of a planet orbiting the Sun is not uniform. The planet moves faster when it is close to the Sun, and slower when it is far away. (c) The time it takes a planet to go around the Sun varies as the (3/2) power of its distance to the Sun. The outer planet in this picture has an orbit four times larger than planet A, and it takes 8 times—that is, 4 to the (3/2) power—longer to go all the way around the Sun. These three laws served as the "experimental proof" of Newton's law of universal gravitation.

that the inverse-square law is the *only* reasonable one that always gives a simple closed curve for an orbit. With either a stronger or a weaker force, the planet will trace complicated rosettes instead. Moreover, Newton's solution is technically valid only for a "solar system" consisting of a sun and *one* planet. If you throw in a second planet, the problem already becomes too difficult to solve exactly. Even if you allow the planet and the sun to deform a little bit so they are no longer perfect spheres, the solution again becomes much more difficult.

Mathematicians and physicists would have to contend with all of those complications in later years. As far as the seventeenth century was concerned, Newton's theory was a spectacular, unqualified success. Newton became a national hero in England, with a knighthood and a comfortable job as master of the royal mint; and when he died in 1727, he was buried in Westminster Abbey.

Rethinking the Beginning

None of the great scientists of the 1600s—Galileo, Kepler, and Newton—ever wrote about the origin of the Moon or the rest of the solar system. In the case of Galileo and Kepler, such restraint is completely understandable. We have already seen how much trouble they got into by speculating on the *present* constitution of the solar system. To delve into its past would have run the risk of even more serious conflicts with the church. It is worth remembering that both Galileo and Kepler were sincere and devout Christians, as well as good scientists. They would never have questioned the biblical account of creation without a compelling reason; and in their era, there was no valid scientific reason to do so.

Newton's reluctance to speculate on origins appears to have been more deliberate. In the 1713 edition of *Principia Mathematica* he wrote: "This most beautiful system of the sun, planets and comets could only proceed from the counsel and dominion of an intelligent and powerful Being." And, more explicitly: "All material things seem to have been composed of the hard and solid Particles above-mentioned, variously associated in the first Creation by the Counsel of an intelligent Agent. . . . And if he did so, it is unphilosophical to seek for any other Origin of the World, or to pretend that it might arise out of a Chaos by the mere laws of Nature."

But Newton could not hold back the tide of speculation for long, and this last quote suggests that he knew exactly where the tide was leading. The idea that the world evolved from a primordial chaos was an old one, dating back at least to the ancient Greeks and particularly the writings of Epicurus. But Epicurus's version provided no physical mechanism that would cause particles in the original chaos to approach one another and stick together. Now physicists had one: the force of gravity.

There was also increasing geological evidence, throughout the 1700s, that Earth could not be as young as the biblical account made it seem. George-Louis Leclerc, the comte de Buffon, a naturalist and keeper of the French national botanical gardens, estimated that Earth was at least seventy-five thousand years old, based on his experiments on the time it takes a ball of molten iron to cool. If Earth was truly this old, perhaps the six days of creation really stood for much longer epochs of time. And if the "days" were only a metaphor, then the parts of Genesis dealing with the formation of Earth, the Moon, and the heavens could also be a metaphor for a more complicated process that could be understood scientifically.

Appropriately, it was also Buffon who proposed what could be called the first "modern" theory of the formation of the planets. In a book published in 1749, he pointed out that all the six known planets (Uranus had not been discovered yet) and all their known moons rotate in the same direction. (Needless to say, Triton, the backward-moving moon of Neptune, also was unknown at this point.) It was staggeringly unlikely that such a coincidence could happen by chance. But, Buffon argued, it could have happened if all the planets were formed at once by the oblique impact of a comet with the Sun. The impact threw off spinning globules of matter, which then cooled to become the planets.

From today's perspective, the most obvious problem with Buffon's theory is that a comet is nowhere nearly large enough to produce such an effect. Sending a comet into the Sun is like dropping a pebble into an ocean. But, of course, Buffon could not have known this. To all appearances, a great comet is every bit as impressive as the Sun, so it is understandable that he thought they were comparable in size.

However, Buffon's theory had other shortcomings that soon became obvious. As the physicist Pierre-Simon de Laplace pointed out, any globules that were dislodged from the Sun would necessarily be placed into a very long, elliptical orbit, quite unlike the nearly circular orbits the planets have today. Even worse, they would make it through only half of one orbit before crashing back into the Sun.

To rescue his theory, Buffon would have needed something to happen to the globules *after* they were detached, and before they crashed again, to round out their orbits and move their perihelions away from the Sun's surface. By the time Laplace published his critique, Buffon had died, so he never got a chance to answer it; it's unlikely that he could have. But it will be worth bearing Laplace's argument in mind when we come, two hundred years later, to a very similar theory for the origin of the Moon.

In 1755, a second version of the origin of the solar system appeared, which got off to a terrible start but contained many ingredients that many scientists still accept.

Immanuel Kant is better known as a philosopher than as a scientist, but his book *Universal Natural History and Theory of the Heavens* would do any scientist proud. In the eighteenth century, the boundaries between disciplines were much more fluid than they are today. Science was only a minor branch of philosophy, called "natural philosophy." There was nothing exceptional about a man starting out, like Kant, as a natural philosopher and then moving into other areas of the subject. Nor was it unusual that the comte de Buffon, though most noted for his work on plants and animals, would also speculate about probability and planetary physics.

Kant's model of the solar system began with an initial, formless cloud of gas or smoke, which contracts under the force of gravity. One might expect the cloud to simply collapse to a point, end of story. But Kant assumed that the "fine particles" in the cloud would also repel each other. (He made an incorrect analogy to the diffusion of smoke, which he thought was due to a repulsive force.) This repulsion would give them a sideways motion and allow them to take up circular orbits around the growing Sun. After many collisions between these orbiting particles, an overall direction of rotation would be established, and something very much like what physicists now call an *accretion disk* would emerge.

Once the particles have set up shop in this disk, the repulsive forces between them mysteriously switched to attractive forces. As Kant noted, "The beginning of the formation of the planets is not to be sought in the Newtonian Attraction alone. This force would, in the case of a particle of such exceptional fineness, be far too slow and feeble. One would rather say that in this space the first formation takes place by the concourse of certain elements which are united by the usual laws of combination, till the mass which has arisen thereby has gradually grown so large that the Newtonian force of attraction in it has become powerful enough to increase it more and more by its action at a distance." This comment could have been written by a scientist today (though perhaps not in such elegant language).

Kant went on to make several other predictions. They would not all prove to be correct, but he did a remarkable job for someone writing in 1755. He predicted planets beyond Saturn, which would eventually merge by "insensible gradations" into comets. The Milky Way, he thought, is formed by the same process that formed the solar system, and thus also has a huge central mass. He conjectured that there are other star systems (the word "galaxy" didn't exist yet) beyond the Milky Way. These form an even grander super galaxy, and the super galaxies are part of a super-duper galaxy, and on and on. . . . "There is here no end but an abyss of a real immensity, in presence of which all the capability of human conception sinks exhausted," he wrote. But Kant painted, on the whole, an inspiring picture, not a depressing one. He saw creation as an ongoing process, with matter organizing itself in an ever-growing sphere of greater and greater perfection, surrounded by an infinite region that is still in its state of primordial chaos.

As far as the Moon is concerned, Kant assumed that the same process happened on a small scale. The Moon formed from a nebula of matter falling into the growing Earth, deflected by repulsive forces. One could interpret this account as the precursor of what later became known as the *coaccretion theory* of the Moon's origin, in which Earth and the Moon evolve together out of the same raw materials. Kant also explained why the Moon always shows the same face to Earth, deducing that its rotation had been slowed down by tidal forces. He argued that Earth's rotation must have been slowed down as well. We will see in the next chapter how George Darwin parlayed this fact into a completely different story of the Moon's origin.

Considering the number of things that Kant got right—accretion disks, planets beyond Saturn, star systems beyond the Milky Way, the tidal locking of the Moon—one might think that his book was enormously influential. Unfortunately, through no fault of his own, it was virtually ignored. At the time, he was still a young, relatively unknown scholar who had just finished his doctorate. The printer went bankrupt soon after the book was published. Almost all the copies of the book were impounded, and were not released until ten or eleven years later. By then, Kant had moved on to other things, and "natural philosophers" had little reason to pay attention to his obscure, hard-to-find treatise. Laplace published a somewhat similar hypothesis in 1796. When the similarities to Kant's theory were pointed out to him later, he claimed to have had no knowledge of Kant's book. It was only in the 1840s that scientists began to give Kant his due and began to refer the "Laplace-Kant Theory."

How to Build a Solar System

Pierre Simon, later the marquis de Laplace, was born in 1749 to a poor family in Normandy. He came to Paris in 1768 and wasted little time making a name for himself; already by 1769, at age twenty, he was a professor of mathematics at the Paris Military School, and by 1773 he was elected to the Royal Academy of Sciences. He apparently was not very popular with his fellow mathematicians. Some found him arrogant, and others claimed he borrowed other people's ideas too freely without giving them credit. Later in his life, they must also have envied his political connections. He had the good fortune to meet Napoleon Bonaparte when the latter, at age sixteen, was taking examinations at the military school where Laplace taught. Laplace continued to cultivate Napoleon's friendship, and when the former sublieutenant Napoleon became Emperor Napoleon, he rewarded Laplace with a seat in the Senate. When Napoleon fell in 1815, Laplace didn't fall with him; he transferred his allegiance to Louis XVIII and was rewarded by being made a marquis.

Laplace published his version of the birth of the solar system in an appendix to a 1796 book called *System of the World*. He begins with a very appropriate note of caution: "I will suggest a hypothesis which appears to me to result with a great degree of probability . . . which however I present with that diffidence, which ought always to attach to whatever is not the result of observation and computation." Like Kant, he proposed that the solar system began as a giant gaseous nebula. Unlike Kant, he was able to use actual observations to support his theory: William Herschel had by then reported the first telescopic observations of nebulas, which appeared to be stars surrounded by a "shining fluid." Some were more concentrated and others more diffuse, and Laplace interpreted these as planetary systems in different stages of development.

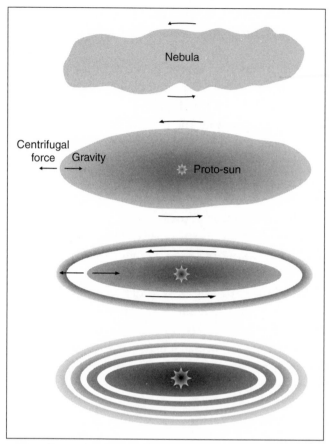

Laplace's nebular hypothesis, proposed in 1796 for the origin of the solar system. As the primordial nebula cools and condenses, it begins spinning faster. At a certain point the centrifugal force on the gas molecules at the outer edge equals the gravitational force pulling them toward the proto-Sun, and these molecules remain at the same distance while the rest of the nebula continues to contract. The process repeats itself several times, and the nebula detaches a series of rings, that eventually condense and form planets. On a smaller scale, this process repeats itself around each planet, thus accounting for moon systems and Saturn's rings. Earth, a small planet, got only one Moon.

In Laplace's version of the "nebular hypothesis," a giant ball of hot gas originally extended as far as the orbit of Uranus (the seventh planet, which Herschel also had discovered recently). As the nebula cooled, it shrank and also began to spin faster, just as a skater spins faster when the skater draws his or her arms in. At some point, the Sun's gravity could not provide enough centripetal force to hold the outermost layer of spinning gas, and this layer would separate off into a ring. Then the nebula would cool off further, shed another ring, and so on. Meanwhile, the rings themselves also would cool off and begin to condense. "If all the particles of a ring of vapors continued to condense without separating," Laplace wrote, "they would at length constitute a solid or liquid ring. But the regularity which this formation requires in all the parts of the ring, and in their cooling, ought to make this phenomenon very rare. Thus the solar system presents but one example of it; that of the rings of Saturn." Instead, the solid or liquid material in the ring would usually cluster together and its self-gravitation would round it into spherical masses. Laplace did not specifically address the origin of the Moon, but the reference to the rings of Saturn makes it clear that he sees this process, as Kant did, taking place on a smaller scale about each of the planets.

Laplace's nebular hypothesis, whether seen as a distinct theory or an elaboration of Kant's, became the dominant paradigm for the next century, so that any subsequent theory of the Moon's origin would have to fit into it. In particular, scientists were convinced that the solar system and Earth began as hot objects: a hot, gaseous nebula and a molten ball of lava, respectively.

Longitudes and the Lunar Theory

In the 1700s and 1800s, as the world economy became increasingly dependent on long-distance sea voyages and exploration of unknown lands, navigation on the open seas became one of the greatest scientific problems of the day. Even in charted waters, a ship could run aground if its course had been plotted incorrectly. And in uncharted waters, what would be the use of discovering a new island or harbor if you might never find it again?

The position of a ship can be described by its latitude and longitude. Finding the latitude—the ship's distance north or south of the equator, measured in degrees—was a fairly routine matter. In the northern hemisphere, it is especially simple. The higher in the sky the North Star is, the farther you are from the equator; for example, if it is fifty degrees above the horizon, you are at fifty degrees north latitude. If it is just on the horizon, you are on the equator.

Longitude was a whole different matter. Nowadays, longitude is measured in terms of the number of degrees east or west of the prime meridian, which passes through the Royal Observatory in Greenwich, England. However, that has been the custom only since 1884; before then, a captain would measure his position with respect to a port in his home country.

The lack of a standard "starting place" for longitudes was only a minor part of the main problem, which was that no one had a good way of measuring east-to-west distances. Because of Earth's rotation, the longitude difference corresponds to the difference between local time and the time at your port of departure, with each hour of time corresponding to fifteen degrees of longitude. The local time is easy to determine, because the sun is highest in the sky at noon. But it is not so easy to keep track of the time in your port of origin. Until 1761, there were no clocks capable of keeping steady time on board a rocking ship. Even as late as the early 1800s, naval chronometers were still too expensive to be standard equipment. Thus sailors had to look for other ways to keep track of the time back home.

For scientists, the longitude problem became a tantalizing target. The Dutch government offered Galileo a chain of gold if he would work on it. The English Parliament, in 1714, offered a staggering twenty thousand pounds sterling for a solution. In 1716, the duke of Orleans offered a hundred thousand francs. Finding the longitude was the eighteenth century's equivalent to sending a man to the Moon, in terms of the money spent on it, and in terms of the perceived unlikelihood of success.

For a long time, until naval chronometers became widely available, the Moon provided the most accurate way of determining the longitude at sea, and mathematicians worked diligently to compute its motion. This led to the birth of a whole other "lunar theory," which referred not to the *origin* of the Moon but to its *motion*. The lunar theory in turn stimulated the entire field of celestial mechanics, the study of the motion of objects in space. Thus we owe our ability to steer spaceships, in part, to the eighteenth-century mathematicians who were trying to figure out how to steer ships at sea.

The Moon entered navigation via the method of *lunar distances*. The Moon's angular distance from the Sun changes by roughly half a degree every hour. A sailor could measure the angular separation between these two bodies with a sextant. (At night, he could measure the separation between the Moon and a reference star instead.) If he found that the Moon and the Sun were 117 degrees and 30 minutes apart, he could look up a table in his *Nautical Almanac* that told him at what time of day or night—at the Royal Observatory in Greenwich—the two bodies were expected to be that distance apart. In this way he could find out the current time in Greenwich and thus figure out his longitude.

This method depends, of course, on accurate forecasts of the Moon's motion. The Moon is the fastest-moving body in the sky as seen from the Earth, and also the hardest to predict. The English government offered a three-thousand-pounds-sterling reward for anyone who could forecast the Moon's motion with sufficient accuracy to compute the longitude to within a degree. This prize was eventually awarded to the Swiss mathematician Leonhard Euler (who developed the theory) and the German mathematician Tobias Mayer (who did the calculations).

Why is the Moon's orbit so hard to predict? Didn't Newton prove that it moves around the Earth in an ellipse? Not exactly. Newton's proof would work only if Earth and the Moon were perfectly round (which they aren't) and the only bodies in the solar system (which they aren't). In fact, the Sun is a veritable King Kong looking over the shoulders of Earth and the Moon. Its pull on the Moon is stronger than Earth's. Instead of calling the Moon a satellite of Earth, we should really call it our celestial roommate. Earth and the Moon are both satellites of the Sun, occupying the same orbit and leapfrogging past each other roughly twelve times a year.

From the very beginning, Newton knew that the Moon's orbit was not a two-body problem but a three-body problem, an intricate minuet choreographed by the Sun, the Moon, and Earth. Try as he might, he could not find an elegant equation for the Moon's orbit; supposedly he told Halley that the problem "made his head ache and kept him awake so often that he would think of it no more." Nor was Newton the first to be frustrated by the Moon. Kepler also was never able to determine what sort of curve the Moon traces out, even though he had conquered Mars. As we have already mentioned, one way in which the Moon departs from a purely Keplerian orbit is the nineteen-year precession of the nodes. This is only one of many aberrations caused by the Sun.

To see how difficult it is to understand how the Sun affects the Moon's motion, here is a little quiz: Does the Sun make the Moon circle Earth more quickly, less quickly, or at the same rate it would if the Sun were not present?

On first thought, it may seem that the Sun makes no difference at all. The Moon moves away from the Sun for half a month (from new Moon to full Moon), so the Sun's pull slows it down for that half; but then it moves back toward the Sun for another half a month (from full Moon to new Moon), during which the Sun speeds it up. The two effects ought to cancel each other out, right?

If you agreed with this, you are guilty of Keplerian thinking. Think again of the stone at the end of the rope. The key factor that controls how quickly it goes around is the tension in the rope—in other words, how quickly the stone is accelerating *toward the center.* Instead of thinking about whether the Sun's pull slows the Moon or speeds it up—which has to do with acceleration *around* the center— we should think about whether the Sun is pulling the Moon *toward* the center of its orbit (i.e., toward Earth) or *away* from the center.

In fact, the Sun tends to slow down the Moon's monthly trip around Earth. At new Moon, when the Moon is between the Sun and Earth, the Sun pulls harder on it than it does on our planet. That makes the Moon move outward slightly, and that means it slows down, like a spinning skater who moves his or her arms outward. At the first quarter, when Earth and the Moon are about the same distance from the Sun, the Sun has very little effect on the Moon's speed. At full Moon, when the Moon is farther from the Sun than Earth, the Sun pulls less strongly on it than Earth. From our Earth-centered point of view, it looks as if the Sun were pulling the Moon outward, again away from Earth. Therefore, just as at new Moon, the Moon moves outward from Earth and slows down. Finally, at the waning quarter, just as at the first quarter, the Sun's influence is minimal. Overall, the Sun slows the Moon's revolution down by roughly one hour a month, which means that each lunar cycle takes one hour longer than it would according to Newton's two-body theory.

So much ado over a silly hour! But it's a very big deal if your estimate of the time in Greenwich is off by an hour. Each hour in the sky corresponds to one *time zone* of difference in longitude—the distance between New York and Chicago. Small wonder that mathematicians of the eighteenth and nineteenth centuries were kept very busy working out this and other effects of the Sun's disturbing force.

To make things worse, each perturbation feeds back on the others; for example, a change in the inclination of the Moon's orbit will affect the rate of precession of the nodes. A complete solution to the Moon's orbit requires a careful accounting for all the primary effects, then a reaccounting for the secondary effects, and so on. The books on celestial mechanics in the 1800s were filled with many pages of formulas of this sort; it was a time of tremendous perspiration and little inspiration, as mathematicians worked doggedly toward a seemingly elusive ideal of perfection.

The tremendous difficulty of the three-body problem of Earth, the Moon, and the Sun leads to another question that bothered Newton: Is the solar system really stable? Even in the case of Earth, the Moon, and the Sun, the perturbations could conceivably conspire to throw the Moon completely out of Earth's gravity—or cause it to spiral in and destroy us. The solar system as a whole is even more complex. Newton thought it was unstable, and that the Creator had to step in now and then to put the planets back in the right places.

In 1675 Edmond Halley made a discovery that gave some credence to Newton's opinion. By studying centuries of observations of Jupiter and Saturn, he concluded that Jupiter was gradually moving toward the Sun and Saturn was gradually spiraling away from it. Clearly, if the trend continued, either God would have to intervene or the solar system would be a very different place in a few million years.

In 1784 Laplace showed that the Deity need not trouble Himself. Jupiter's inward drift was only a temporary one, caused by the near-resonance of its orbit with Saturn: Jupiter completes five orbits of the Sun in very nearly the time that Saturn completes two. This means that they come closest to each other over and over again at roughly the same spot in their orbits—which is exactly when they perturb each other's orbits most strongly. But Laplace showed that they would eventually perturb each other in the other direction. In fact, Jupiter and Saturn repeat a cycle of spiraling in and spiraling out about once every 850 years.

More than this, Laplace proved (considering only primary effects and neglecting feedbacks) that the solar system *is* stable: none of the planets are undergoing long-term perturbations that would eventually throw them on a collision course, or cast them out of the solar system entirely. All the apparent long-term trends, such as Jupiter's spiral toward the Sun, are part of even longer-term oscillations; hence they will eventually reverse themselves.

It was a very optimistic view of the solar system, and very much in keeping with the scientific mood of the time. Nature worked according to rational rules, which the human mind could decipher. In geology, scientists were moving away from a cataclysmic history of Earth to a uniformitarian history, with no biblical flood. Similarly, Laplace's solar system was uniformitarian, with no violent events, and could take care of itself without God's intervention. When Napoleon questioned why Laplace's great opus *Celestial Mechanics* did not mention God, Laplace replied succinctly and truthfully, "Because I had no need of that hypothesis."

5

Daughter Moon

"I'm bursting with delight at my work of the last few days," wrote George Darwin to his father, Charles Darwin, on April 25, 1876. "I've been thinking day and night over Evans's suggested problem about the alteration in the axis of the earth. After long thought I have got through the hardest part and I think I see exactly what is the mathematical problem involved. . . . I'm rather counting my chickens before they're hatched, but I'm regular [*sic*] bursting with ideas on the subject."

Little did he realize it, but George Darwin had just stumbled on his life's work. His feverish ideas of that April would turn into a series of papers about Earth's rotation, about the tidal interactions between Earth and the Moon, and finally about the distant past of Earth and the Moon. Although a few people had jotted down some ideas about the Moon's origin, no one that we know of before Darwin really thought of it as a separate problem from the formation of the rest of the planets. And certainly no one subjected his or her ideas to the intense mathematical analysis, running to thousands of pages of calculations, that Darwin ultimately did. If Galileo and Kepler brought the Moon down to Earth (Galileo through his telescope and Kepler through his imagination), then Darwin was the man who brought its past into the present.

Britain's First Family of Science

Few people were ever brought up in a time or milieu more conducive to learning than George Darwin. Born in 1845 in Down, County Kent, England, he was fourteen years old when his father's monumental work *The Origin of Species* was published. His father enrolled him in a new style of public school, one where the sciences were emphasized as much as philosophy, religion, and the classics. Charles

Darwin never pushed his five sons, but he did support any interests that arose naturally. When the young George became fascinated by optics, Charles bought him a set of lenses. Charles also taught him the line from Anthony Trollope that became George's personal motto: "It's dogged as does it."

Sundays were a highlight of life at Down. Charles Darwin did not go out very often in later life; instead, friends and fellow scientists came to visit him. The long hours of conversation on the scientific and political topics of the day made for "the most agreeable society I have ever known," wrote George. His father was seldom eloquent ("I do not know that I have ever heard anyone whose sentences so often contained some infraction of grammatical rule," George recalled), yet always to the point.

With such a childhood environment, it is not surprising that all of Darwin's sons grew up to have impressive careers. The oldest, William, became a banker. Francis, the third son, followed in his father's footsteps and became a naturalist, and ultimately—like both Charles and George—a knight of the British Commonwealth. Leonard, the fourth, followed Charles in a sadder way. He became a leading eugenicist, trying to use the principles of evolution to prove the superiority of certain classes or races over others. It was a respectable theory at the time that became infinitely less respectable after World War II. Horace, the fifth and last son, founded the Cambridge Scientific Instruments Company, for many years a leading supplier of laboratory apparatus, and served briefly as mayor of Cambridge.

George was not an overnight success. He went to Cambridge University and graduated in 1868, taking second place in the university's vaunted mathematics competition, the Tripos. But after that, he vacillated, dithered, and struggled with poor health, which would become a lifelong annoyance to him. He studied for the bar but did not go into a law practice; he tried his hand at sociology, statistics, and economics but failed to accomplish anything of note.

Then, toward the end of 1875, he struck gold. One of the biggest geological puzzles of that time was—and still is—the cause of ice ages. A fashionable theory, generally advocated by biologists who knew little about physics, attributed them to shifts in Earth's axis. In other words, Earth's poles had moved around over millennia, causing glaciers to appear in places that don't have them anymore. One scientist had suggested that this movement could have been caused by the rising of new continents out of the sea. Darwin didn't believe it, and in April 1876, the time of the creative frenzy he described to his father, he sat down to prove it.

Darwin considered an extreme scenario: How far would the pole shift if a continent a quarter the size of the Northern Hemisphere suddenly emerged from the ocean? He worked out that each foot of rise would move the pole by fifteen yards, so even a rise of ten thousand feet—the height of the Tibetan Plateau, highest on Earth—would move the pole by only one and a half degrees. There was no way this theory could explain ice ages. Darwin wrote a paper on the subject and, feeling "like a pea meeting a cannonball," discussed it with Britain's leading physicist, Sir William Thomson (later Lord Kelvin). Thomson praised the paper, and Darwin was on his way to a new career. With that, Thomson became a lifelong friend.

Meanwhile, the ideas continued to bubble forth from Darwin's pen. If a change of axis couldn't explain ice ages, perhaps a change in the obliquity of Earth's rotation could. This would be different from the other theory: The North and South Poles would remain in the same geographic locations on Earth, but the tilt of the axis of rotation with respect to the ecliptic plane would change. The Earth's present twenty-three-degree tilt has a powerful effect on climate; it is responsible for the seasons, and it delineates the Arctic and Antarctic Circles. George Darwin's idea was that tides—not tides in the ocean, but distortions in the entire shape of Earth—had caused the tilt to grow.

By 1878 George had written a paper proving his thesis about the tides, and his stock was rapidly on the rise. On October 29 of that year his father wrote him a letter awash in paternal pride: "My dear old George, I have been quite delighted with your letter and read it all with eagerness. . . . Hurrah for the bowels of the earth and their viscosity and for the moon and for all the heavenly bodies and my son George (F.R.S. very soon)." As he wrote these words, Charles Darwin's usual cramped handwriting grew larger and larger and darker and darker, and the last three words were a great, ungainly scrawl of emotion.

Charles forecasted his son's triumph correctly. (He may have had inside information.) George was indeed made a Fellow of the Royal Society (F.R.S.) in 1879, which, for a British scientist, was the equivalent of earning a union card. Four years later he became a professor

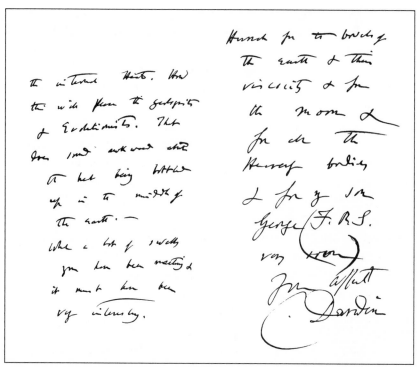

When George Darwin's first papers on tidal friction came out, his father, the world-famous naturalist Charles Darwin, could scarcely contain his pride. "Hurrah for the bowels of the earth and their viscosity and for the moon and for all the heavenly bodies and for my son George (F. R. S. very soon)," he wrote in this letter from October 29, 1878. *Courtesy of Cambridge University Library.*

of astronomy at Cambridge University, where he would spend the rest of his life. By then, unfortunately, Charles was no longer alive to cheer him on; the elder Darwin had died in 1882.

The Fission Theory

Darwin's theory of the origin of the Moon grew out of his work on the tides; he first began to work on it in about 1878, and he continued to add details throughout his life. It is a theory worthy of the son of the great evolutionist, because it is very much an *evolutionary* theory of the Moon—and Darwin himself thought of it in those terms.

Darwin began by investigating the tidal forces of the Moon on Earth, which he had already shown would lead to a gradual increase in the tilt of Earth's axis. Everybody knows that the Moon creates tides in the oceans, but the same forces also must create bulges in the solid body of Earth, though they are not as noticeable as the ocean tides. If Earth had a molten beginning, as all astronomers thought at the time, then the tides might have been even stronger at that stage.

Very roughly speaking—perhaps *too* roughly—the Moon makes Earth bulge in two places: the part directly beneath the Moon, and the part directly opposite the Moon. But because Earth rotates faster (once a day) than the Moon revolves around it (once a month), its rotation carries the bulges past where they would otherwise be. This has to do with the viscosity of Earth, as noted in Charles Darwin's letter to George; the rocks and oceans cannot relax immediately once they are past the region of greatest tidal force. One might compare Earth to a giant glob of silly putty, which only grudgingly changes its shape when you squeeze and knead it.

So we can visualize Earth as a spherical body with two bulges: one closer to and a little bit ahead of the Moon, the other farther from and a little behind the Moon. The Moon's gravity tends to pull the nearer bulge backward, because it is behind that bulge. For the same reason, the Moon pulls the more distant bulge forward. But because the gravitational force on the nearer bulge is stronger, the overall torque is backward, against the direction of Earth's rotation. The Moon is gradually slowing Earth's rotation, making the day get longer.

This much had already been figured out by Kant, although he waffled on it later in life. Like many other scientists, Kant believed that Earth had contracted as it cooled, and that the contraction would speed the day, thereby counteracting the tidal effect.

By Newton's first law ("Every action has an equal and opposite reaction"), if the Moon was pulling backward on the tidal bulges, those bulges must be pulling *forward* on the Moon. It would be natural to assume that this makes the Moon orbit Earth faster, but that would once again be a Keplerian/Cartesian mistake. Thinking again of the rock swinging around at the end of a rope, if the rock starts moving faster, it will tend to pull out on the rope. Unless you resist that force, the rock will move outward, and the radius of the circle it moves in will increase (assuming there's more rope). In space, of course, there is no rope holding Earth and the Moon together, so the Moon *does* move outward. The net effect of the tidal bulges is not to make the Moon go faster, but to make it retreat from Earth. As it

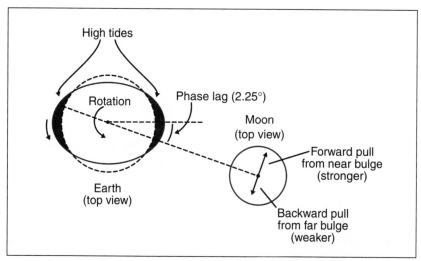

George Darwin's theory of tidal friction. Earth's rotation, which is faster than the Moon's orbital motion, carries Earth's tidal bulges around too fast for the Moon to keep up. As a result, the nearer high-tide bulge is slightly ahead of the Moon, and the farther one is slightly behind. However, the gravitational force from the nearer bulge is slightly stronger, so there is a net forward pull on the Moon (in addition to the much larger pull toward Earth's center, which is not shown here).

moves away, the month gets longer (because of Kepler's third law). Thus both the month and the day are getting longer. But the mathematics revealed that the day is lengthening more rapidly than the month is, so at some time in the far future it will catch up. At that time, Darwin worked out, the month and the day will both last fifty-five of our present days. Earth will always show the same face to the Moon, just as the Moon already shows the same face to Earth, and this arrangement will be stable for all time.

What if we run the movie backward? As we move backward in time, the Moon gets closer and closer to Earth. Both the day and the month get shorter and shorter, until the Moon is whizzing around Earth once every five hours and thirty-six minutes, and Earth is rotating at the same speed. At this point, the Moon's surface would loom over Earth, only five thousand miles away. To put it another way, a flight from Earth to the Moon would be shorter than a flight from London to San Francisco. But that is assuming that both bodies retained their present shapes. If they were molten, they would be hugely distorted by the immense tidal forces on them. The Moon would have a cigar shape, perhaps twice as long as it was wide. The two bodies would be practically touching one another.

At this point, the movie reel, frustratingly, runs out. The physics gives no valid clues for inferring what happened earlier, because the situation just described is also an equilibrium. But anyone watching this movie would have little doubt about what happened next: the Moon would merge with Earth. Of course, that's in the "backward" version of the movie. Running the movie forward again, we would see Earth flinging the Moon off into space. And that (minus the movie-projector imagery) is how George Darwin came to propose his fission theory of the Moon's birth. "These results point strongly to the conclusion that if the Moon and Earth were ever molten viscous masses, then they once formed parts of a common mass," he wrote in his seminal paper, published in 1879.

A Matter of Timing

Based as it was on quantifiable physical effects, Darwin's theory soon achieved wide acceptance among scientists. For the wider public, it was probably reminiscent of all the old myths in which the Moon Goddess is born from the Earth Goddess. Mother Earth in Darwin's theory is truly a mother, and the Moon is her daughter planet. In 1882 a geologist named Osmond Fisher suggested a further detail that made the analogy even more compelling: perhaps the Pacific Ocean was actually the birth scar left by the separation of the Moon. Although Darwin never adopted this idea in his own writings, it soon became part of the popularized version of his theory. The combined version, the Moon separating from Earth and leaving behind the Pacific basin, continued to be taught in schools long past the time when scientists had lost faith in it.

From the beginning, there were two major problems with the fission theory, but neither one seemed to be insurmountable. The first was the issue of time. Darwin was remarkably precise in his calculations of the radius and period of the early Moon's orbit, because these could be extrapolated from known data about the Earth-Moon system. But it was much more difficult to pin down the amount of time it would have taken for the Moon to evolve from that original orbit to its present one. That depended on the efficiency of Earth's response to the tide-generating forces. Just how viscous were the bowels of Earth? And how much had their viscosity changed over the years? No one really had any idea. Darwin attempted to deduce

the rate of change of the length of the month by looking at historical records, and from this he came up with a very rough guess. The
Moon must have taken at least fifty-four million years to reach its
present orbit—and that was assuming the most optimistic estimates
of the efficiency of the tides.

This calculation ran smack into a long-simmering controversy
about the age of Earth. At the center of the controversy were Darwin's father and his friend and supporter Lord Kelvin.

In the first edition of *The Origin of Species,* Charles Darwin had
attempted to estimate the length of geological time by computing
how long it would have taken the sea to erode a valley called the
Weald in southern England. He came up with a figure of three hundred million years. It was a very primitive estimate, intended only to
give a rough idea of the time scale of evolution, but critics immediately leaped on it as being much too large. Darwin removed this passage before the third edition of his book appeared, not wanting to
jeopardize the acceptance of his theory of natural selection because
of a minor detail he was not sure of. But over the years, other geologists arrived at estimates in the same ballpark, about a hundred million years, by comparing the thickness of rock strata with an estimate of how rapidly sediments are laid down on the ocean floor.

Balderdash, said Kelvin. Or, in his words, "A complete misinterpretation of the physical laws." He went back to Buffon's idea of fig-
uring out how long it had taken Earth to cool from its molten state.
He was the perfect person to do it: he was an expert on thermody-
namics, and had invented the temperature scale named after him, in
which zero degrees corresponds to absolute zero. He also had the
benefit of new experimental data that showed that rock melts at a
lower temperature than expected, about 1,200 degrees Celsius (2,200
degrees Fahrenheit or 1,500 degrees Kelvin). The cooling time of
Earth, if it started at that temperature, was roughly twenty million to
forty million years, and Kelvin preferred the lower value. Not only
did this conflict with geological evidence, but it also conflicted with
George Darwin's minimum estimate for the age of the Moon.

Fortunately for Darwin (as well as for many biologists and geologists), a totally unexpected discovery unraveled Kelvin's theory of the
age of Earth. First, in 1896, Henri Becquerel discovered the radioactivity of uranium. In 1903, Marie and Pierre Curie discovered radium,
an element that was so much more radioactive than uranium that it
actually gave off heat. (Uranium does, too, but not enough to notice

in a small sample.) Suddenly physicists had a brand-new explanation for the geothermal heat that warms the interior of Earth. It is not remanent heat from Earth's molten past; it is being generated right now by the decay of radioactive elements—principally uranium and thorium—in Earth's interior. Darwin himself was one of the first scientists to point that this invalidated Kelvin's age estimates, which were based on the assumption that no new heat is being generated within the Earth.

Not only did radioactivity expose Kelvin's mistake, it also provided a new and far more accurate way of telling how old Earth really was. At the Cavendish Laboratory in Cambridge, Ernest Rutherford showed that radioactive elements decay at a constant rate, unaffected by any chemical processes. The most common isotope of uranium, for example, has a half-life of 4.47 billion years. This means that a rock that was formed 4.47 billion years ago would have exactly half of its original uranium left; the remainder would have decayed into a number of by-products, principally lead.

Unfortunately, no one can look at a rock and tell what its original concentration of uranium was. But physicists can compare the amount of uranium to the amount of by-products. For example, a rock that had an equal number of uranium atoms and lead atoms would be about 4.47 billion years old. As it turns out, this approach is still a little too naive (in part because uranium decomposes into different forms of lead, and one has to separate the "radiogenic" lead from the primitive lead that had never been through radioactive decay), but it gives a general idea of how physicists can estimate the age of a rock. By 1911, the year before Darwin's death, a mineralogist named Arthur Holmes had found rocks more than 1.6 billion years old. Suddenly Darwin's Moon had all the time in the world to reach its present orbit.

The Pear-Shaped Figure

The second, and more serious, problem with the fission theory was the unexplained gap between a single proto-Earth and the equilibrium configuration Darwin had found, with Earth and the Moon separated by five thousand miles. In essence, the whole "gestation" and "birth" of the Moon were missing.

Darwin did his best to fill in this gap; it became his obsession for thirty years. His daughter Gwen Raverat, in her memoir *Period Piece,* wrote about her childhood image of Darwin, sitting by the fireplace, propping up his red-stockinged feet on the fender (all Darwins, she said, had cold feet) with papers strewn all around him, working on his "pear-shaped figure of equilibrium." Raverat was no scientist, and still less so as a child; one surmises that the pear-shaped figure of equilibrium must have been a common topic of conversation, perhaps even a bit of a joke, in the Darwin household.

The key question Darwin was trying to answer was how a rotating ball of fluid could spontaneously separate itself into two pieces. By his time, mathematicians had already discovered several equilibrium shapes for a rotating ball of fluid held together by its own gravity. Newton started the subject off, in his *Principia Mathematica,* by showing that such a globe would be flatter at the poles and bulge outward at the equator. He did this by imagining two giant boreholes filled with liquid, one from the North Pole to the center of the Earth and one from the equator to the center of the Earth. If the Earth was in equilibrium, the pressure at the center of the Earth would be the same in each borehole; otherwise the liquid would flow out of one and into the other. But the weight of the liquid in the equatorial borehole would be reduced by the centrifugal force from Earth's rotation; to compensate, therefore, this borehole had to be longer than the other one, so it would hold more liquid.

In 1742, the Scottish mathematician Colin Maclaurin worked out the exact shape of a molten, spinning Earth—a sphere flattened into a beanbag shape—and for nearly a century that seemed to settle matters. But in 1834, a German mathematician named Carl Jacobi discovered that a fluid planet does not have to be rotationally symmetric. This is a case where ordinary intuition fails. It simply turns out that if you spin molten Earth fast enough, to the point where the polar radius becomes less than 0.5827 (or about seven-twelfths) of the equatorial radius, Earth is no longer comfortable with a beanbag shape and will spontaneously jump into a slightly squashed cigar shape instead.

This obviously looked like the first step toward giving birth to the Moon. First Earth flattens, then it lengthens. What could have happened next?

In 1885 the French mathematician Henri Poincaré found that a new instability emerges when the rotating cigar shape becomes 0.3451 (or roughly one-third) as high as it is long. His calculations suggested that the new equilibrium figure that emerges is *pear-shaped*, with a large bulge at one end and a smaller bulge at the other. What a huge shot in the arm for Darwin's theory! If Jacobi's ellipsoid was like the first trimester of a pregnancy, with no external signs of an impending birth, Poincaré's pear-shaped figure suggested the second trimester, when the baby starts to show. But it was only suggestive. Unlike Maclaurin and Jacobi, Poincaré did not work out the exact shape. His method indicated only the direction of change at the moment of instability. To use a contemporary analogy, it was like watching a computer "morph" one image into another but stopping after just a few frames, and then trying to guess what the final image might be.

Poincaré was a mathematician, and more interested in general principles than the details of the Earth-Moon evolution. He couldn't be bothered with the drudgery of finding the exact pear-shaped figure of equilibrium. But Darwin, living up to his motto "It's dogged as does it," spent years working on it. He never did finish, and probably with good reason. Later work by the Russian mathematician Aleksandr Lyapunov showed that the pear-shaped figure is most likely unstable. This was not bad news for the fission theory, because it meant that the pear shape would disintegrate quickly, once achieved. But it was bad news for Darwin's project of subjecting the entire fission process to rational analysis.

Nevertheless, in Darwin's lifetime at least his theory had a good deal of mathematical plausibility. As proto-Earth contracted from its hot nebular state, it would have spun faster and faster, progressing through the beanbag shape, the cigar shape, and the pear shape until finally the two ends of the pear separated into Earth and the Moon. Darwin, the son of the world's most famous evolutionary theorist, was delighted by this model, with its abrupt changes of shape so reminiscent of the emergence of species. A less scrupulous scientist might have declared victory. But there was just one more point that made him uneasy.

Darwin knew that the Moon, just after its birth, had to be orbiting once every five and a half hours, and Earth had to be rotating at the same rate. No matter how complicated and obscure the birth process was, it could not have a great effect on the combined angular momentum of the pair. So Earth before the Moon's birth was probably rotating once every three to five hours. That is pretty fast by today's standards, but not fast enough to get Earth past the cigar stage, let alone make it split in two.

If the Moon couldn't have a natural birth, then perhaps its birth was assisted by the Sun. Darwin, as the world's leading authority on tides, thought that the *solar tide* could do the trick. In the present era we think of tides as a lunar phenomenon, but in fact the Sun raises tides on Earth, too. The solar tide-generating force is about half as strong as the Moon's, so its effect now is only to make the lunar tide stronger (a spring tide) or weaker (a neap tide). But before the Moon was born, the solar tides would have been the only ones.

If the solar tidal force is only half as strong as today's lunar tidal force—which is itself only one ten-millionth as strong as Earth's gravity—how could it possibly be strong enough to tear two planets apart? The answer, Darwin suggested, might be resonance. This is the same phenomenon that makes it possible for a musical note, sung at just the right pitch, to break a glass. If the singer causes the air to vibrate at just the same frequency that the glass prefers to vibrate at on its own—called its natural frequency—the vibrations in the glass will grow larger and larger until they break it. Thus a tiny periodic force (the vibration of the air molecules) can be amplified by resonance to produce large effects.

If the solar tides happened to coincide with the natural frequency of the *whole Earth*—which Darwin's friend Lord Kelvin had calculated at one hour and thirty-four minutes—then they could be amplified to the point where Earth, like the glass, would break apart. The tides went through two cycles a day then, as they do now. If the day were three hours and eight minutes long, which was just within the range that Darwin accepted as a possibility, then the tides would go up and down twice in that time—in other words, every hour and thirty-four minutes.

Let us imagine, then, how Darwin's scenario would look from space. The molten Earth spins faster and faster as it contracts. It flattens out into a beanbag, and then possibly a cigar. We would then notice little ripples on the surface of the cigar—the solar tides. The ripples would get steadily larger, amplified by resonance, until the entire Earth would be convulsed in giant birth contractions, pulsating back and forth from one end of the "cigar" to the other. All this time, remember, Earth would be spinning furiously around its shortest axis. Then, suddenly, one end would break off in a shower of droplets, and the Moon would be born.

Even Darwin recognized that he was asking a lot from the solar tides, and he took pains to distinguish this part of the theory from the subsequent evolution of the Moon's orbit around Earth. "There is nothing to tell us whether this theory affords the true explanation of the birth of the moon, and I say that it is only a wild speculation, incapable of verification," he wrote. "But the truth or falsity of this speculation does not militate against the acceptance of the general theory of tidal friction, which, standing on the firm basis of mechanical necessity, throws much light on the history of the earth and the moon, and correlates the lengths of our present day and month."

And that is pretty much how posterity has judged the matter. Astronomers have no doubt that tidal friction is causing the Moon's orbit to grow. The lunar ranging experiments left by *Apollos 11, 14,* and *15* have spoken the last word on the subject: the Moon is receding from Earth at a rate of 3.82 centimeters (about an inch and a half) per year.

The fission theory of the Moon's birth, though, fell out of favor after one of its chief supporters abandoned it in 1930. Harold Jeffreys, a well-known British astronomer, wrote a paper that year showing that the solar tides could not have grown large enough to cause Earth to split; the molten Earth would have been too viscous (i.e., sludgy) for that to happen. After that, there was little debate over the fission theory for the next thirty years. On the eve of the Apollo missions, it was still under a cloud of suspicion but definitely not refuted, and it was widely considered to be one of three plausible scenarios for the Moon's birth. In 1969, a survey of twenty lunar scientists by sociologist Ian Mitroff found that fission was the least popular of the three theories. Nevertheless, three of twenty evaluated it as the most plausible.

Selective Memory

How many people today remember that Charles Darwin's second son was also a leading scientist, "the most eminent of applied mathematicians since Kelvin," as one obituary described him? His obscurity gives away the end of our story, to a certain extent. Had he been proved right—and promptly—our school science textbooks would probably mention two Darwins today.

And yet later events had nothing to do with the quality of his work. He had a good idea, and pushed it as far toward its conclusion as anyone could in his day. He wrote a book on the tides in 1898 that still makes for fascinating and informative reading. He developed a method for predicting high and low tides, called the harmonic analysis of tides, that is still used. (In fairness, it should be said that Kelvin invented the method and Darwin made it workable.) Darwin was in constant demand as a speaker, and in the final year of his life he organized the premier mathematics conference in the world, the International Conference of Mathematicians. Of all the "loonies" in this book, his résumé is one of the most impressive.

Perhaps it is just as well that Darwin never knew the verdict, or the selective memory, of history. He died in 1912, before Europe was ripped apart by World War I. He lived in a genteel era when people like him were known as "men of science," not "scientists" ("a new and mongrel word he abhorred and never used," according to his daughter Margaret Keynes). He was very much Old Cambridge, the Cambridge of the nineteenth century. As his obituary in the campus newspaper read: "The portrait of our subject which would carry with it to some former mathematical student the fullest assurance of the eternal unchangeableness of Cambridge would place Professor Sir George Darwin in his study. He would be seated in a low armchair and a baize-colored board would rest across the arms. Sheets of figures and calculations would be before and around him, and some result would have just been reached which was going to help in probing one of the secrets of the universe."

6

Captive Moon

With his theory of tidal friction, George Darwin had identified the most important phenomenon governing the long-term evolution of Earth's and the Moon's orbits. He had traced the past of the two bodies as far as tidal friction could take him: back to an epoch when Earth and the Moon were whirling around each other like the tigers in an Indian folktale who chased each other around a tree until they melted into butter.

Unfortunately, to get from the whirling-tiger phase to the pool-of-melted-butter phase, Darwin had to take a leap of faith. In spite of his and Poincaré's best mathematical efforts, it remained a big leap. The problem, in technical terms, was that the configuration Darwin had found, with the Moon and Earth tidally locked to each other in a 5.6-hour orbit, was an "unstable equilibrium."

An unstable equilibrium is something like a mound on a putting green in golf. The top of the mound is the unstable equilibrium point. If you putt the ball near to this point, it becomes very difficult to predict or control what will happen next. If you have aimed just slightly to the left of the highest point, the ball will swerve to the left. Just slightly to the right, and it will roll down the hill to the right. If your aim has been perfect, it might roll straight down the other side. Or, if you struck it just a little bit too softly, it might roll back down the hill straight at you. It is just as hard to "retrodict" where the ball came from as it is to predict where it will go. If you see it coming down the hill toward you, you can't tell whether someone hit it from the other side of the mound, or whether you hit it yourself.

To describe what happened to the Moon before it reached its unstable equilibrium with Earth, Darwin made the simplest assumption: If you kept going backward in time, the Moon would keep getting closer to Earth, until eventually the two coalesced. But for the Moon as for the golf ball, other past histories were possible.

Over the years, two rival lunar histories emerged as equally likely candidates. One was the "capture theory," which postulated that the Moon was a separate planet that had formed elsewhere in the solar system and then was lured into Earth's gravitational embrace. The other was the "coaccretion theory," in which the Moon and Earth simply grew up together and have always orbited one another, though they were much closer in the past.

The problem with this celestial whodunit was not so much that all three theories were equally good, but that all of them were equally bad. Until the launch of the Apollo missions, each theory required its supporters to take some sort of leap of faith. It even became a standard joke in the field that lunar scientists were much better at explaining why the Moon *wasn't* there than explaining why it was. "There is no existing theory of the origin of the moon which gives a satisfactory explanation of the earth-moon system as we now know it," wrote Ralph Baldwin, possibly the leading lunar observer of the 1940s and 1950s. "Somehow ... the two bodies were formed and became partners. But how?"

The Capture Theory

If George Darwin was the essence of the nineteenth-century British gentleman, Thomas Jefferson Jackson See was nineteenth-century America: brash, self-made, and self-promoting.

Born in Missouri in 1866, See was named after two heroes of the Confederacy: Jefferson Davis, its president, and Stonewall Jackson, one of its greatest generals. He had an ego and a rebellious streak worthy of his name. To See, there was no problem too difficult for him to solve, no theory too outrageous for him to advocate, and no accolade so great that he couldn't apply it to himself. The title of a book that he probably instigated and certainly collaborated on says it all: *The Biography and Unparalleled Discoveries of T. J. J. See.* The nominal author of the book, William Webb, calls See "the greatest astronomer in the world, and one of the greatest of all time." Against this glowing praise we may set the opinion of the president of the University of Missouri, See's alma mater, who wrote to See's first employer that the young graduate was "thoroughly unscrupulous, an intriguer among students of a dangerous type, a genius in prosecuting his own claims to preferment, and to sum up, largely devoid of moral principle." And this professor was writing about his university's top student.

See grew up in rural Missouri, where the prairie grass grew "as high as a horse's back," in Webb's (or maybe See's) words, and he graduated first in a class of thirteen students at the University of Missouri. While he was there, See gained two years of experience on the school's $7\frac{1}{2}$-inch telescope. That convinced him he wanted to become an astronomer, and in the fall of 1889 he set sail for Germany, where he enrolled as a graduate student at the University of Berlin.

It is hard to imagine what kind of impression this educated farm boy must have made in Kaiser Wilhelm's Germany, with his dashing handlebar mustache, his towering stature (six feet, four inches) and his broad shoulders. But his unwavering self-confidence stood him in good stead; he assiduously sought the acquaintance of Europe's leading scientists, including George Darwin. He was already intrigued by the origin of the planets, but for the time being he heeded the advice of respected astronomers such as the American Simon Newcomb, who warned him, "In the present state of our knowledge, such investigations lead to nothing."

Instead, See made his reputation as an observer of double stars. These are pairs of stars that orbit in close proximity to one another, and they were one of the hottest topics in nineteenth-century astronomy. They intrigued theorists because of their intense gravitational and tidal interactions, which provided a good test for Newton's theory of gravitation. See discovered more than four hundred of them in his lifetime, making him one of the champion double-star finders of all time.

After he received his doctorate in 1892, See sailed back to America and the opportunity of a lifetime. The oil tycoon John D. Rockefeller had just founded a new university in Chicago, and the university's "boy president," William Rainey Harper, was rounding up top scholars from all over the country to teach at the prestigious new school. The dynamic Harper had already persuaded a wealthy Chicago businessman, Charles Yerkes, to fund a new observatory that would have the world's largest refracting telescope (a telescope that uses lenses rather than mirrors). It was, in fact, an era of grand new observatories: the Lick Observatory in California was built in 1888, the Lowell in Arizona in 1894, the Yerkes in Wisconsin (because the urban environment of Chicago was deemed unsuitable) in 1897, and the Mount Wilson in California in 1904. By being present at the founding of one of these great observatories, See could have set himself up for a lifetime. But he squandered his chance through personal jealousy.

It clearly grated on See that he was the second fiddle in the Astronomy Department. Just before he hired See, Harper had brought in George Ellery Hale to oversee the planned observatory and head the Astronomy Department. In fairness to See, it was an awkward position. Hale's father was a wealthy businessman, who had financed a private observatory for his son (the Kenwood Observatory), which Hale used until the forty-inch telescope at Yerkes was ready. Hale wouldn't even let See touch the Kenwood telescope, because he considered him incompetent. For his part, See considered Hale an incompetent manager, and apparently offered Harper some unsolicited advice as to how he could save the university $15,000 on the operation of the Yerkes Observatory.

To make things worse, in 1895 See sent Harper a remarkable twenty-two-page ultimatum in which he demanded a promotion from instructor (the lowest rank on the faculty) to associate professor (the same rank as Hale). In fact, See strongly implied that he deserved more: "I am in every respect entitled to the position of [full] Professor of Astronomy," he wrote, but he magnanimously offered to accept the title of associate to spare the university any embarrassment. It's clear that Harper was beginning to regret hiring this *enfant terrible,* but he did offer See a promotion to assistant professor (still a step beneath Hale). That wasn't good enough for See, and he left in a huff. According to his biographer Webb, "at the University of Chicago nothing was being done on merit."

It was a story that was repeated with depressing regularity throughout See's career. No matter where he went, people found his arrogance insufferable—and his science increasingly sloppy. At his next stop, the Lowell Observatory, See lasted only two years before he was fired. A controversy with a former student of his, Forest Ray Moulton, who had pointed out some mistakes in an article See wrote for the *Astronomical Journal,* led to his being banned from publishing in that magazine. He got a job at the U.S. Naval Observatory in

Washington, D.C., which had an excellent twenty-six-inch refractor—but this opportunity soured, too. After a bout of "stomach trouble" and an unsuccessful semester as a mathematics instructor at Annapolis, he finally landed in 1903 in the astronomer's equivalent of Siberia: an observatory at the naval base on Mare Island, near San Francisco. There, instead of the state-of-the-art telescopes he was used to, he would be working with a five-inch refractor, smaller than the one he had used as an undergraduate in Missouri. He would live in the officers' quarters (right next to the cows' stables) on an island base that had no bridge joining it to the mainland. His main job would be to keep the standard time for the West Coast, drop a time ball at noon each day to allow the ships in the naval yard to synchronize their clocks, and send out a series of "time ticks" over the Western Union wires from 11:55 A.M. to noon.

Keeping the standard time for the entire West Coast was a significant responsibility, but it was a job far beneath See's qualifications. Undoubtedly his superiors in the navy—not to mention his fellow astronomers—watched him move to California with at least some sense of relief.

If so, they were wrong. Lacking the resources to do meaningful telescopic research, See decided to become a theorist instead. He returned to the enigma of the origin of the solar system, which Newcomb had warned him not to waste his time on. Faced with increasing difficulty in publishing his work in reputable science journals, See found that he could get his work out just as well by sending his findings to the bewildered editors of local and national newspapers. Impressed by his credentials and unaware of the controversy he had stirred up, these would often publish See's "press releases" lock, stock, and barrel.

Thus, on June 26, 1909, See's new theory of the Moon's origin was announced not in a scientific journal but in newspapers such as the *San Francisco Call* and the *New York Times*. The *Call* reported that "Professor See . . . declared that he had demonstrated mathematically that the nebular hypothesis of La Place [*sic*], the groundwork of astronomers for more than one hundred years, is unscientific and untrue." The *Times* was somewhat more reserved—"He rejects entirely the long-accepted theories of Laplace and Sir George Darwin ascribing earthly origin to the moon"—but did devote eight full paragraphs to a detailed explanation of See's capture theory.

Captain Thomas Jefferson Jackson See (1866–1962), posing in his office on Mare Island. One of the most controversial figures in American astronomy, See was the first prominent scientist to propose the "capture theory" of the Moon's origin. *Photograph courtesy of Vallejo Naval and Historical Museum.*

As usual, See couldn't resist embroidering the truth a little for the sake of the newspapers. The *Call* describes a "lecture" given by See at a meeting of the Astronomical Society of the Pacific. In fact, the president of that society, William Wallace Campbell, had specifically denied See's request to give a public lecture at the society's meeting, on the grounds that the subject matter was "extremely technical." Instead, Campbell agreed to grant him a private ten-to-fifteen-minute audience with the Board of Directors. In all likelihood (as Campbell's letters show no indication that he changed his mind), this was how the capture theory really made its debut.

What is supposed to matter in science, of course, is not who proposes a theory or where, but how they defend it. The theory is supposed to be judged on its merit. As much as the Astronomical Society of the Pacific would have liked to sweep See's theory under the rug, he did have substantive arguments. See used current and accepted research on the dynamics of the "three-body problem"—in other words, the way that three objects (in this case, the Sun, Earth, and the Moon) move if the only force on them is Newton's inverse-square law of gravity. He later detailed his argument in an enormous

750-page self-published book, *Evolution of the Stellar Systems, vol. II, The Capture Theory*.

In a nutshell, See argued that space is not a vacuum. It is, or at least was, filled with very tenuous matter or a "resisting medium." It was by the repeated braking action of this matter over many encounters that the Moon, which formed independently of Earth, gradually had its energy depleted until Earth could capture it. And See didn't limit his theory to the Moon; he concluded that all the planets had been captured by the Sun in the same way. The action of the resisting medium, he felt, was as fundamental a force in celestial mechanics as Newton's law of gravity.

If See's idea sounds a little bit far-fetched, bear in mind that ninety years later, NASA would use exactly this method, called "aerobraking," to slow down the Mars Observer spacecraft, place it into orbit about Mars, and then round out that orbit into a circle. What See did, to the extent possible with nineteenth-century mathematics and no computers, was to show that aerobraking was possible.

There was no problem with See's mathematics, which in any case were mostly borrowed from George Darwin and George William Hill, an American mathematician. The problem was with his assumptions, as the directors of the Astronomical Society of the Pacific undoubtedly informed him. First, Hill's work, which showed that the two larger bodies in the three-body problem would set up zones in which the smallest body could be trapped, was valid only if the third body was tiny compared to the first two. In the case of the Mars Observer spacecraft, this assumption is certainly true. But the Moon, at one-eightieth of Earth's mass, is not exactly negligible. To put it somewhat differently, asking a tenuous resisting medium to slow down the Moon was like trying to slow down a train with a barrier of tissue paper.

See was well aware of this objection, and must have argued that there was enough time over the aeons for the interplanetary medium to do its job. But there was another hole in his system, one so huge and glaring as to disqualify the whole thing. What was this "resisting medium," anyway? Certainly not Earth's atmosphere, which reaches up only a few miles.

In his book See never bothers to address this question. He simply says, in truly oracular fashion, that the resisting medium is there. In an explanatory letter to Campbell, See said that the resisting medium *formerly* pervaded the solar system. Thus it could have been Laplace's nebula, or it could have been a swarm of planetesimals, as See's nemesis Forest Ray Moulton had proposed. This would have been the most respectable answer scientifically, but it was one that See's ego could hardly accept. Remember that he wanted to *refute* Laplace, and certainly he would not have wanted to refute him by accepting Moulton's model instead.

Over the years, See tried out various other explanations. His theory of the resisting medium bore a vague resemblance to the "luminiferous aether" that had been proposed in the nineteenth century as the medium through which light waves propagated. That theory had, however, been discredited by the famous Michelson-Morley experiment that failed to detect any such aether. Instead, See said that his resisting medium actually consisted of egg-shaped particles of light. A few years later he decided that it was the luminiferous aether after all. The rest of the scientific world was just wrong.

If See had expected the capture theory to restore his reputation, it didn't happen. Over the last fifty years of his life—he lived to be ninety-six, and died in 1962—he kept on writing books that no serious physicist or astronomer paid any attention to, proposing new theories about everything under the Sun, from the cause of earthquakes to the origin of comets to the nature of gravity and magnetism. Every now and then he got something right. *The Capture Theory* volume contains a splendid chapter on the formation of the Moon's craters by impacts. See was one of the first scientists to accept this idea, which was considered wacky at the time. He illustrated it with pictures of the craters formed by bullets striking a lead target at a naval firing range, which were very similar in shape to Moon craters.

Unfortunately, in his later life See's notoriety came mostly from a vitriolic public campaign he led against Albert Einstein, whom he accused of being a fraud and a plagiarist. To his credit, Einstein never joined in the mudslinging, but other scientists were forced to do so on his behalf. An example of See's poison pen, and a depressing sign of his xenophobia, was a letter he wrote to Robert Aitken, Campbell's successor at the Lick Observatory: "It has seemed to many persons very strange that a great American Observatory had nothing better to do than to spend its breath defending the indefensible [i.e., Einstein]," See wrote. "Is not such a course thoroughly un-American and unmanly? Can Americans of self-respect tolerate such running after the unsound reasoning of discredited foreigners, to the neglect of recognized discoveries of the most lasting character, made in our own country?" He was undoubtedly referring to his own theories.

But the greatest victim of See's malice was surely himself. In his exile on Mare Island, ignored by his scientific peers, he must have been desperately lonely. Perhaps his delusions of grandeur kept him from realizing how much had he lost; perhaps he was content with his life as Captain See, U.S. Navy, and his increasingly outdated responsibilities as timekeeper for the West Coast. But the man who could have been one of America's first prominent astronomers ended his life as nothing more than the village eccentric, "the sage of Mare Island." He was the crazy old guy that the *Vallejo Evening Chronicle* always sent its cub reporters to interview, knowing they would return with their heads spinning from an incomprehensible two-hour rant. His was a first-class mind shackled to a monstrous ego, and the ego won.

A Tidal Mating Dance

The capture theory, however, fared much better than its promulgator. Particularly in the 1950s and '60s, it was probably the most thoroughly studied of the three theories of lunar origin. In Ian Mitroff's survey of lunar scientists on the eve of lunar exploration, mentioned earlier, five of twenty rated it as the most likely scenario of the Moon's formation. But it is also worth mentioning that a modified version of the capture hypothesis, called the many-moons theory, received six votes in Mitroff's survey. If the two versions are combined, capture hypotheses were actually the favorite of the majority of the experts.

In any capture hypothesis, the biggest problem is how to slow the Moon down enough to bring it into Earth orbit—or, to put it in physicists' language, how to dissipate enough of its energy. You can think of the Moon as a giant bungee jumper leaping toward Earth from a platform somewhere out in space, infinitely far away. If the bungee cord works perfectly, the Moon will snap right back to infinity after its close approach. But a real bungee jumper dissipates energy in many ways—for example, through internal friction within the cord, or external friction with the surrounding air. This means that the bungee jumper never snaps back all the way to the place he or she started from.

See had identified one possible way of dissipating energy: a resisting medium in space. This would be like the air that resists the motion of the bungee jumper. However, he could not convince other scientists that such a medium actually exists today. It could have existed in the past, in the form of debris orbiting Earth soon after its formation. Unfortunately, it would take almost a Moon's worth of planetesimals to slow the Moon down sufficiently, and in that case, scientists would be right back where they started. They would have to explain where all those planetesimals had come from.

But there was a second obvious possibility, one familiar to physicists ever since George Darwin had identified it back in the nineteenth century: tidal friction. Only this time it was the tides raised in the *Moon* that would do the trick. Just like the bungee cord, the Moon would be stretched by the tidal force during its close encounter with Earth, and if it was not perfectly elastic, it would lose energy. It would not be able to "snap back" all the way to the place where it originated, and thus Earth would be able to pull it back again. Over repeated encounters, the Moon's apogee (its farthest distance from Earth) would decrease and its perigee (nearest distance to Earth) would increase, until its orbit became nearly circular, as it is today.

You may be skeptical that tides would be strong enough to do this. But in fact, tidal forces are weak only if the body that exerts the force is very far away, as Earth's Moon is distant from Earth today. At close range tidal forces can easily tear a satellite apart, unless it has some internal forces holding it together. For a spacecraft or an astronaut, they pose no threat; the chemical bonds inside a solid are much stronger than the tidal forces. But a newly formed moon would probably be nothing more than a pile of rubble or a ball of liquid, and thus it would be very vulnerable to tidal effects.

The most vivid example of capture and of tidal breakup unfortunately took place a century after George Darwin, too late for it to have any effect on the debate on the Moon's origin. This was the spectacular demise of Comet Shoemaker-Levy in 1994. This comet was captured by Jupiter sometime before 1992 (astronomers didn't see it happen). On its last pass around Jupiter in 1992, tidal forces tore the comet apart into more than twenty pieces; they also altered the orbit enough that the comet entered a collision course with the planet. That was the point at which Carolyn and Gene Shoemaker and David Levy discovered it, in 1993; fifteen months later, the whole world saw its fiery end.

But even before Comet Shoemaker-Levy, there was one other excellent example of tidal forces in the solar system, namely the rings of Saturn. In 1850, a French astronomer named Édouard Roche showed that the outer limit of Saturn's rings falls at almost exactly the limit of stability for a moon: about 2.46 times Saturn's radius. Within that radius, the tidal forces from Saturn prevent any particles within the rings from ever coalescing into a moon. The same would have held true for Earth's Moon as well, with one slight modification: because the Moon is less dense than Earth, the Moon's self-gravitation is weaker in comparison to the tidal force, and it would break up farther out, about 2.9 Earth radii instead of 2.46. This "Roche limit" is one of the most important constraints in all theories of the Moon's origin. Though the theories themselves have come and gone, it is quite impressive that the Roche limit has not needed any significant revision since 1850.

For the capture of the Moon to work, it must have passed close enough to Earth to experience a lot of tidal disruption, but not so close that it fell apart like Comet Shoemaker-Levy or the rings of Saturn. The first physicist to work out how this might have happened was a German high-school teacher named Horst Gerstenkorn, who wrote a series of papers on the capture model from 1955 to 1969. Later, other physicists, such as Gordon MacDonald of England, would join in, and the advent of computers in the early 1960s made it possible to test various versions of the capture hypothesis.

It is important to realize that computer simulations of the capture hypothesis—or computer simulations of any theory of the Moon's origin, for that matter—are not really designed to pin down a unique history. Instead, they are designed to test a range of possible beginnings for the Earth-Moon system, and hopefully identify features common to all of them. If the scenario is too unique—for example, if the capture can take place only under very precise and unlikely circumstances—it actually works against the theory.

Viewed in this light, Gerstenkorn's and MacDonald's work proved to be both a boon and a liability for the capture theory. They clarified many details that could not have been anticipated without a careful mathematical analysis, but at the same time they made the capture look like a very unusual event. Here is how they believed the mating dance of Earth and the Moon went:

- In any reasonable capture scenario, the Moon had to penetrate inside the Roche limit. But Gerstenkorn argued that this would not cause it to disintegrate because it would venture inside for only an hour or two out of each orbit, and would spend a much longer time outside the Roche limit. He called this a "diving" orbit, because the Moon would dive inside the danger zone and then escape again.
- The actual capture would take place on the Moon's first pass, which would be so close that it would bend the Moon's trajectory by ninety degrees or more. This right-angled turn would take place in fewer than two minutes!
- After this, the Moon would continue for a while in a very elongated orbit. It would take five to fifteen years for the orbit to round out enough that the Moon would no longer venture inside the Roche limit. After five thousand years the distance between Earth and the Moon would be ten Earth radii (compared to sixty today), and the orbit would be essentially circular. This was the point at which Gerstenkorn stopped his simulations.
- In many simulations the Moon would actually be traveling in a retrograde orbit (east to west) when it was first captured. Its orbit would then flip over the North and South Poles into the present prograde direction, just like a top flipping over.
- The intense energy dissipation during this time would heat the Moon throughout by 2,000 to 5,000 degrees Centigrade (3,600 to 9,000 degrees Fahrenheit). This would guarantee that the newly captured Moon was molten.
- A very crucial discovery was that the Moon could be captured only if it was moving at most 5 percent faster or slower than Earth—like a car moving sixty-three miles per hour that overtakes another one moving sixty miles per hour. If the difference in velocities was any greater, the only way to get enough tidal slowing would be for the Moon to run into Earth! If Gerstenkorn had pursued this idea, he might have discovered the giant impact hypothesis. However, impacts were anathema in that era, and Gerstenkorn always treated this as a possibility strictly to be avoided.

The last point convinced many scientists that capture was very unlikely. Kepler's third law implies that if the Moon was traveling at almost the same speed as Earth, the Moon also had to be traveling in nearly the same orbit. Basically, the Moon had to be steered through a very narrow window at a very slow speed. Even Gerstenkorn realized that this was a problem.

To compound the problem, he and other lunar scientists were working with an incorrect age for the Moon. This age had been derived using Darwin's model for tidal friction, assuming that Earth's tidal bulges were off center by the same amount they are today, roughly 2¼ degrees. There really was no good physical reason for thinking that this "phase lag" would stay constant over the aeons; if the oceans changed their configuration (as we know they have), then the phase lag would probably change as well. But in 1963, biologist John Wells discovered evidence that the phase lag has been approximately constant throughout biological history. From the analysis of rings in fossilized coral, he deduced that the length of the day in the Devonian period (four hundred million years ago) was about twenty-two hours. This was the same length predicted by Darwin's theory, assuming a constant phase lag. It really was rather weak evidence, considering the uncertainty in the count of the growth lines in the coral. Also, though four hundred million years is a long time in the *biological* history of Earth, it is less than a tenth the age of the planet; the phase lag could have changed radically in earlier times.

Still, when weak evidence is the only evidence you have, you go with it. So, extrapolating the tidal friction backward, physicists computed that the Moon had its close encounter with Earth 1.4 billion to 1.6 billion years ago. And that was bad news for the capture theory. Because the solar system is a little over 4.5 billion years old (a figure that was accepted in the early 1950s and is still accepted today), the Moon somehow had to go around the Sun, in an orbit very close to Earth's, for 3 billion years *without* being captured or ramming into Earth. This simply couldn't be done. So the Moon had to have come from somewhere else, perhaps the asteroid belt. But this, too, made no sense. Gerstenkorn himself had shown that Earth could not capture the Moon in one fell swoop from an orbit that started in the asteroid belt, because the Moon would be traveling too fast. So something had to throw the Moon out of the asteroid belt, and something else had to alter its orbit to match Earth's, all so it could enter the narrow window of velocity that would allow it to be captured by Earth. It was too great a chain of miracles. As Gerstenkorn himself laconically wrote, "It remains obscure how the planetoid ever could enter this orbit."

However, capture theorists did not give up. The most famous of all lunar scientists at the dawn of the space age, Harold Urey (about whom more later), believed that the solar system once consisted of numerous Moon-sized objects, and that the Moon was a last relic of that era. It was very important to him—more as an article of faith than of rational conviction—that the Moon had *never* been melted, so it had retained a record of the early solar system. He viewed capture as a gentler process than fission (although Gerstenkorn's model would seem to suggest otherwise) and thus favored the capture hypothesis. To counter the argument that capture was very improbable because of the narrow window of acceptable velocities, he would say that there were many such encounters while Earth was forming. Many of the Moon-size planetoids got away. Others—eighty of them, judging from Earth's total mass—got absorbed into Earth. And one of them, the Moon, came along with just the right velocity to go into orbit about Earth. With enough tries, even a long-shot gamble will pay off.

Urey's argument required a certain amount of willful blindness to the results of Gerstenkorn's simulations. In 1962 Thomas Gold (another well-known lunar maverick, most famous for his warning that any spacecraft that landed on the Moon would be swallowed up by a deep layer of dust) suggested a new variation. In Gold's theory, which was worked out by Gordon MacDonald in more detail from 1962 to 1964, Earth had captured several smaller moonlets—according to MacDonald, six to ten of them. These different-sized moonlets would move out from Earth at different rates due to tidal friction and thus they would gradually accumulate, over the course of three billion years, into one Moon.

The many-moons theory had points to make everyone happy. It explained why the Moon formed so late in Earth's history. The capture of the moonlets would not be quite as improbable or as high-energy a process as the capture of a complete Moon. In fact, Mars has two "moonlets" even today, Phobos and Deimos, that are almost certainly captured asteroids. Because it was such a recent idea, the many-moons revision of the capture hypothesis was not as well known when Ian Mitroff conducted his interviews of lunar scientists in 1969. It is revealing that only six of the twenty scientists he interviewed expressed an opinion about it, but all six of them considered it the most likely theory of the Moon's origin.

It also would be one of the first to go after the Apollo missions.

7

Sister Moon

In the debate over the Moon's origin, the third and final popular theory before the Apollo missions was the coaccretion theory, which postulated that the Moon and Earth grew up at the same time and in the same part of the solar system. In this theory, the creation of our Moon was not a unique or dramatic event. The Moon formed by the same process as all the other planets and the larger moons— although opinions changed over the years as to what that process might be.

The coaccretion theory was different from the fission and capture theories in another significant way: it lacked a high-profile advocate in the English-speaking world. The fission theory had Sir George Darwin, who is largely forgotten today but in his time was one of the world's foremost physicists, as well as a scion of the greatest family in British science. The capture theory had Thomas Jefferson Jackson See, who brought notoriety to the theory, if not respectability. Later, and more importantly, it had the Nobel Prize–winning chemist Harold Urey, whose belief in a "cold Moon" (see chapter 8) led him to accept capture as the most likely scenario.

The leading advocates of a Moon and Earth that grew up together came first from France and later from Russia. Because of the Cold War barriers between the United States and Soviet Union, the Russian work on coaccretion remained unknown in the West until the very eve of the Apollo missions. It is tempting to speculate on what might have happened in a parallel universe where the Russians beat the Americans to the Moon. Perhaps American lunar science would have withered on the vine, deprived of the gusher of public support and government money that allowed it to flourish in the 1960s and early 1970s. Perhaps the theory of coaccretion, which was dominant in Russia, would have remained unchallenged as the "correct" view of the Moon's origin.

The Man Who Never Laughed

As soon as astronomers discovered other moon systems—first around Jupiter, then around Saturn, Uranus, and Neptune—they interpreted them as solar systems in miniature. As we have already seen, this analogy was the decisive factor in causing Galileo to come out of the closet and admit his belief in the Copernican theory. The moons of Jupiter demonstrated that the same principles were operating everywhere in the solar system, the principles that Newton later elucidated with his universal law of gravitation.

If the moon systems and the solar system as a whole obeyed the same physical laws, then it made sense that they should have the same origin. In Laplace's theory, the solar nebula had once extended as far as the most distant present-day planet (which in Laplace's time was Uranus), and then shed rings of gas as it cooled and contracted. These rings eventually coalesced into hot balls of gas. Then the process repeated itself: The planetary nebulas acted just as the solar nebula had, condensing into a planet in the center and shedding moons in their outer extremities.

However, it was Laplace who first noticed that Earth's Moon posed a challenge for this theory: it seemed to be much too large and far away to have formed out of the proto-Earth's nebula. If the nebular hypothesis was right, the proto-Earth should have extended all the way out to the Moon's present distance and probably a good deal beyond, to shed a ring large enough to condense into the Moon. Yet any gas molecule at such a great distance from Earth—sixty times the radius of Earth—would have been more strongly attracted to the Sun than to Earth. Therefore the Sun would have torn apart Earth's nebula had it been so big. At best, Laplace calculated, Earth's nebula could have extended out to only forty Earth radii.

George Darwin's theory of tidal friction would have given Laplace an easy way out of this paradox if he had known about it: The Moon could have formed inside forty Earth radii and then retreated to its present distance. But Darwin's tidal theory didn't come out until 1879. The first person to rescue the coaccretion idea was actually a countryman of Laplace, in 1873.

Édouard Roche, an astronomer and mathematician who lived almost his entire life in Montpellier, on the Mediterranean coast of

France, was in some ways a French version of George Darwin. He came from a well-respected bourgeois family, to which he remained closely attached throughout his life. Like Darwin, he had what was called in those days a "delicate constitution." However, unlike Darwin, who traveled far and wide in search of health, Roche was content to remain in Montpellier. He was described by his contemporaries as modest, even timid; for example, he turned down three offers to become the dean of the University of Montpellier. (Perhaps he was simply showing good sense.) His modesty did not prevent him from getting due recognition for his work; he was made a knight of the Legion of Honor in 1863.

Born in 1820, Roche showed great promise as a teenager and would have gone to study in Paris, but his health prevented it. He taught himself at home with singular success—reading through Laplace's formidable opus *Celestial Mechanics,* making some of the first measurements of solar prominences during the total eclipse of 1842, and writing a dissertation on the shape of the planets in 1844 that earned him a doctorate from the University of Montpellier.

At that point Roche moved away for the only time in his life and spent three years hobnobbing with the scientific elite at the Sorbonne in Paris. In 1847 he returned to Montpellier to get married. Unfortunately, this joyous occasion turned into a shocking tragedy. His bride suddenly became violently ill during the festivities, and eight days later she died of an "inflammation of the chest." Roche was a changed man after that. "It was from that day, I believe, that he ceased being able to laugh," wrote his former student M. J. Boussinesq, "though keeping from habit that amiable smile, a somewhat melancholy expression of his kindness, that all of his students have known." Roche never remarried, but threw himself instead into the pursuit of science.

At that endeavor he was much more fortunate. Earlier, I mentioned his discovery of the "Roche limit," the distance of about 2.4 Earth radii (depending slightly on density) inside of which any liquid or gaseous satellite will be torn apart by tidal forces. He also was way ahead of his time in deducing that Earth has an iron core. At the time most geologists believed, thanks to the intimidating Lord Kelvin in England, that the interior of Earth was molten and homogeneous. The present idea of a differentiated Earth, with a stony mantle and an iron core, emerged very gradually through analysis of earthquake traces, and did not become the orthodoxy until after 1910. But

Roche derived the existence of a core in 1881, without any seismological data at all. His clues were Earth's density; the amount of flattening at the poles; and the precession of the equinoxes, which gave him an estimate of Earth's "moment of inertia," a way of measuring its resistance to turning. It came out differently from a fluid Earth of uniform composition, but agreed very well with a two-part Earth, with the outside made of stone and the core made out of iron. Ironically, this discovery would one day turn out to be a critical piece of information against Roche's own theory of coaccretion.

Roche sketched out that theory in 1873, when he pointed out a rather simple mistake that Laplace had made. In Laplace's theory, the proto-Earth and the proto-Sun were engaged in a tug-of-war over the gases in proto-Earth's nebula. On one side of the tug-of-war, the gravitational self-attraction of proto-Earth pulled the gases toward its center. On the other side, two forces—the centrifugal force due to the rotation of proto-Earth and the force of the Sun's gravity—pulled the gases outward, away from the center. The outer limit of the nebula was set by the point where those opposing forces canceled each other out.

But Laplace had given an unfair handicap to the Sun's side in the tug-of-war. The Sun's gravity pulled on proto-Earth just as hard as it did on the surrounding gases, and thus it could never pull them *apart*. The real tug-of-war was between Earth's gravitation and the Sun's *tidal* force, which could separate gases from proto-Earth. Once again, tides were the key to the Moon's formation—this time, the tides created in proto-Earth by the Sun. But the tidal force from the Sun was relatively weak, and consequently Roche calculated that anything within 237 times Earth's radius would remain within Earth's embrace. Thus the hot ball of gases that formed Earth could easily have been large enough to give birth to the Moon as well.

Still, Roche's work did not completely vindicate Laplace's theory. Even if the Sun's tidal force could not win the cosmic tug-of-war, it still would have stretched the gas cloud out into a watermelon shape. But Laplace's theory of the detaching rings was very particular: it would not work for anything but a circular nebula. If proto-Earth were elongated, then the two long ends of the watermelon would be the most vulnerable to the Sun's tidal force. As proto-Earth cooled and contracted, it would spit out fireballs from the ends of the "watermelon" instead of shedding rings of gas. It would be easy to form small moonlets this way, but not a complete Moon.

To resolve this difficulty, Roche assumed that the Moon had begun as a slightly cooler region in the contracting nebula. The gas would begin to condense there, and the increased gravitational attraction of this region would draw more material in. Eventually, as the surrounding hot gas cloud contracted, the cooler region would work its way to the surface. Suddenly, like a rock exposed by the receding tide, the proto-Moon would emerge from proto-Earth and separate from it. The nebula would, of course, go on contracting and become Earth.

Unfortunately, Roche gave absolutely no explanation of how this giant cool spot could have formed in the first place. Up to that point he had stood on relatively solid mathematical and physical ground. He had found that Earth's nebula could have reached beyond the Moon's current orbit. He had realized also that the nebula could not have shed rings. However, instead of concluding that the Moon had not formed this way, he resorted to making up a "just-so story" that was not based on mathematics at all.

Roche was not the first or the last cosmologist to make this mistake; it seems to be a common failing of aspiring world-builders. And it's not even necessarily a mistake. It's always possible that one of the just-so stories will eventually turn out to be right. A cool spot *could* form in a hot nebula; one only needs to look at Jupiter's giant red spot (which is colder than the surrounding atmosphere, and has presumably been that way for centuries or millennia) to see a rough analogue. But scientists hate flukes. If possible, they like their theories to be based on hard evidence and indisputable chains of cause and effect. Sometimes, especially in the study of the distant past, they have to settle for less. Even then, though, they at least like to know how probable the event is that they are accepting. Roche's theory gave no clue.

How to Build a Solar System (Part Two)

A more important barrier to the acceptance of Roche's coaccretion theory, though, was that Laplace's nebular hypothesis was already on its way out. To physicists, it had a major Achilles' heel: the present-day Sun is turning too slowly to account for its having thrown off such large planets in the past. But for people who believed their eyes instead of mathematical calculations, there was also visual evidence that something was wrong with the nebular hypothesis. For decades,

The Andromeda galaxy, formerly known as the Andromeda nebula. The first photographs of Andromeda and other spiral galaxies appeared in the 1880s and 1890s. These "nebulas" were widely supposed to be planetary systems in formation, and their spiral structure seemed to deal a fatal blow to Laplace's theory, which predicted rings. Only in the 1920s did astronomers realize that Andromeda is not a fledgling planetary system— it is a giant, fully developed star system, like our Milky Way. *Photograph courtesy of Jason Ware, GalaxyPhoto.com.*

its supporters had claimed that nebulas—diffuse glowing patches in the night sky, such the Andromeda nebula, that do not look like pinpoints of light—were examples of solar systems in formation. But with the advent of astrophotography, astronomers got a better look at the structure of nebulas, because a photographic plate "sees" differently from a human eye. It can sit in place for hours, gathering more and more light so that even the faintest features can emerge.

In 1888 Isaac Roberts, an amateur astronomer from Liverpool, England, took a photograph of the Andromeda nebula that revealed for the first time its gorgeous, sweeping, spiral arms. Over the next decade, James Keeler of the Lick Observatory in California amassed many more pictures of spiral nebulas, proving that the Andromeda nebula was not an exception. The sky was filled with spirals, but not a single ring-shaped nebula could be found. If these were solar systems in formation, then the planets must have formed out of spiral arms, not rings.

The first serious competition to Laplace's nebular hypothesis came from an unexpected quarter—not from an astronomer, but a geologist. Thomas Chrowder Chamberlin was another of the scientific luminaries whom William Rainey Harper had lured to the University of Chicago with John D. Rockefeller's money. Chamberlin had started out to be a paleontologist, but because he had settled in Wisconsin, which is full of glacial terrain, he sensibly became a glaciologist instead. Like George Darwin, he started thinking about what caused ice ages, and he was one of the first scientists to realize the importance of carbon dioxide as a "greenhouse gas" that controlled the climate. In Chamberlin's view, the level of carbon dioxide in the atmosphere rose and fell over time, but these oscillations were merely deviations from a *steady* average concentration. But this did not square with the reigning theory of a gradually cooling Earth. If early Earth had been hot, it would have had a denser carbon dioxide atmosphere, and there would have been a dramatic drop in carbon dioxide after the oceans formed. The carbon dioxide would have dissolved into the water and would eventually be removed permanently from the atmosphere by being locked up in carbonate rocks. Since he did not believe this carbon dioxide depletion had occurred, he concluded something must be wrong with the hypothesis that Earth had started out hot.

Chamberlin, unlike See, realized that he needed help when venturing outside his own discipline, and at the University of Chicago he found just the ally he needed: Forest Ray Moulton, See's erstwhile student who later became his nemesis. Between 1898 and 1904 the two of them worked out a new theory of the origin of the solar system that they dubbed the planetesimal hypothesis. The distinguished, white-bearded Chamberlin provided the geological intuition, while the younger Moulton provided the expertise in celestial mechanics and the mathematical computations.

The planetesimal hypothesis had two main points to it, which philosophically are quite separate, although at the time they seemed closely intertwined. First, Chamberlin proposed that Earth and the other planets were assembled from *cold* material—stones and dust instead of a gaseous nebula. "I urged very specifically . . . that the molten globe idea and all that is a misinterpretation, and that Earth is solid, has always been solid, has grown up as a mass of little solid particles, and that these have worked upon each other," he told an interviewer in 1928. It was a brave challenge not only to the long-dead Laplace, but to the very much alive Lord Kelvin in England, who called it a "very sure assumption" that Earth had started out hot and whose authority none had dared to contradict before Chamberlin.

Second, Chamberlin and Moulton proposed that this cold material had been drawn out of the Sun in two long spiral jets by a close encounter between the Sun and another star. The jets had been hot initially, of course, but cooled rapidly into solid fragments long before they began accreting into planets. This part of the hypothesis, clearly, was strongly influenced by the photographs of nebulas with their prominent spiral arms. It may seem surprising that the encounter between the Sun and the other star would produce *two* arms, but that is the nature of tides. (Yes, once again tides enter the discussion. This time they are the tides in the Sun produced by the passing star.) The star, assumed to be larger than the Sun, would stretch the Sun out in two directions—toward the star and away from it—and cause it to eject blobs of matter from both ends. The Sun's rotation would distort the jets into a spiral.

The close-encounter idea remained fairly popular until the 1920s; it was endorsed by prominent astronomers such as James Jeans and Harold Jeffreys. But in 1923 Edwin Hubble knocked the foundation out from under it when he discovered that the so-called spiral nebulas were really galaxies, completely outside the Milky Way galaxy. The Andromeda galaxy, for example, was five hundred times farther away than had been previously believed, and therefore five hundred times larger. It was not a solar system in formation, nor were any of the other spiral galaxies; they were much too large. They *contained* solar systems. So there was not a shred of evidence, after all, that our solar system had begun this way.

By the mid-1940s, the pendulum had swung back again, and a nebular origin of the solar system was back in fashion. However, the planetesimal theory of Moulton and Chamberlin did make one permanent change in the landscape. Cosmologists accepted the idea now that the solar nebula must have cooled off *before* planet formation got started. The planets began as little bits of rock and dust, not as immense clouds of gas. Planets are formed by addition, not by subtraction.

Coaccretion Reappraised

Winston Churchill once described the Soviet Union as "a riddle wrapped in a mystery inside an enigma." During the 1960s and 1970s, this especially described the Soviet space program. Everything about it was top secret: the place where flights were launched (Baikonur), the complex where astronauts trained (Star City, only an hour's drive from Moscow), and even the name of the chief designer of the Soviet rockets (Sergei Korolev). There was nothing secret, though, about the Otto Schmidt Institute of Physics of the Earth, a small branch of the Academy of Sciences in Moscow where most of the Russian research on the origin of the Moon was done. Yet for all the attention that Western scientists paid to it, the Schmidt Institute could as well have been situated on the far side of the Moon.

Otto Schmidt was a larger-than-life character in Soviet science, a geologist who organized expeditions to the Arctic, edited the *Great Soviet Encyclopedia,* and received the country's highest award, Hero of the Soviet Union, in 1937. In the 1940s he became interested in the origin of the solar system and founded the institute that would later bear his name. The institute began small, with only half a dozen scientists. One of them was a young astronomer named Victor Safronov, who had just finished his dissertation for a candidate's degree (equivalent to a doctorate in the West) in 1948.

Schmidt pioneered an approach to the formation of the planets that would prove to be enormously successful. "Schmidt emphasized that the problem of formation of the planets in the already existing solar nebula is relatively independent of the problem of formation of the nebula and should be studied without waiting for the solution of the latter," wrote Safronov, together with his wife and collaborator Evgeniya Ruskol, in 1994. Like Chamberlin, Schmidt was convinced that Earth had never been melted but had accumulated from solid planetesimals.

It was Safronov, though, who really figured out how accretion works and how the solar system got from a disk with millions of planetesimals to one with nine planets. First, he showed that the disk becomes gravitationally unstable and begins clumping up when it passes a certain density. (In the West this phenomenon is called the "Goldreich-Ward instability," but Safronov discovered it first.) After individual clumps reach a diameter of a hundred miles or so, a new stage of "runaway accretion" takes place. Until that stage, two objects would only collide when their paths happened to intersect. But a hundred-mile-wide clump of rock is massive enough to start sucking in objects that would otherwise have missed it. Its effective size, thanks to gravity, becomes larger than its material size. Thus the first planetesimal to reach this size gains a huge advantage over its neighbors. It starts vacuuming up everything in sight. In this way, nine of the planetesimals won and became planets. Their relative sizes were more or less governed by the size of the "feeding zones" (another of Safronov's concepts) that they drew material from.

What made Safronov's version of the solar system utterly different from any previous version was simply this: it was ruled by impacts. The solar systems of Laplace, Roche, Darwin, See, and Chamberlin were shaped by gradual processes—Newtonian dynamics, tidal forces, the dynamics of rotating fluids. Impacts inevitably bring an element of chance into the solar system. No one can predict how billiard balls will move when a player breaks the rack, and no one can predict which planetesimal will win in the game of accretion.

Safronov was the first scientist to recognize that some of the features of our present-day solar system are entirely due to chance. Why is Uranus tipped over on its side, so that its axis of rotation lies almost in the ecliptic plane? Why is Earth tipped by twenty-three degrees? Why does Venus rotate very slowly, and in a retrograde (east to west) direction? All of these things are the result of chance. To be more precise, Safronov assumed that all the planets have a certain "systematic" direction and speed of rotation, which is altered in a random way by impacts. The largest impacts have by far the greatest effect: It would take *a million* ten-mile-wide asteroids to produce the same change in Earth's rotation as *one* hundred-mile-wide asteroid, because the smaller impacts tend to cancel each other out, while the single large impact has nothing to cancel it out. Thus the orientation of a planet's axis and its speed of revolution are strong clues about the most dramatic collision in its past. The one on Uranus must have been a doozy: Safronov estimated that it had been hit by a planet roughly the size of Earth. Earth, he thought, given its tilt, had probably been hit by an object a tenth the size of the Moon (or a thousandth the size of Earth).

One counterintuitive thing about randomness is that it can be, in the long run, very predictable. That is why casinos make money.

Similarly, in the roulette game of the early solar system, there were some things Safronov could predict. He worked out how long it would take for Earth to accrete from the dust ring: about a hundred million years. He also showed that if you took a census of the planetesimals in a feeding zone, you could predict how many planetesimals of each size you would find. This distribution of sizes, called a "power law," governs everything from the size of asteroids in today's asteroid belt to the size of craters on the Moon. It would be a key ingredient in the thinking of the American astronomer Bill Hartmann when he formulated the impact theory of the Moon's origin in the early 1970s.

Odd as it may seem, Safronov himself did not take his theory to its logical conclusion. Perhaps it was the influence of his mentor Otto Schmidt, who believed that Earth had not melted in the course of its formation; this imposed an upper limit on the size of the impacts Earth could be subjected to. Or maybe it just didn't occur to him that the existence of the Moon, not the tilt of Earth's axis, was a record of the biggest event in Earth's history. For whatever reason, he remained firmly in the coaccretion camp, along with his wife, one of Otto Schmidt's last students, Evgeniya Ruskol.

In Ruskol's and Safronov's model, a swarm of planetesimals once surrounded Earth out to 250 times Earth's present radius. (This number is quite similar to the size of Roche's gaseous nebula.) Unstable by itself, this swarm would have been constantly replenished by the material attracted toward Earth from Earth's feeding zone. Finally, when the material in the feeding zone had been exhausted, the swarm collapsed into a ring and then coagulated into the Moon.

Just like the fission theory and the capture theory, the coaccretion theory made sense to some people but seemed highly improbable to others. For starters, Safronov's simulations of accretion were extremely simplistic. They had to be, because they were carried out with nothing more than pen and paper. He did not have access to a computer in the 1950s and 1960s, so he worked everything out by pure mathematical reasoning. It was an amazing feat, and a tribute to the unaided power of the human mind. But to make the problem accessible to mathematics, he had to make simplifying assumptions—for example, that all the planetesimals in the feeding zone were moving at the same velocity. The accretion times he came up with also turned out to be too long. For Neptune and Uranus they came out to be longer than the age of the solar system—according to his models, those planets don't exist yet. He acknowledged that some other processes, which were not understood, must be at work in the formation of the outer planets. For Earth, though, his estimate of a hundred million years still seems reasonably accurate.

These fully understandable shortcomings certainly did not invalidate the idea of simultaneous development of the Moon and Earth. However, there were two objections that were more difficult to answer. One was the problem of angular momentum. We have already seen that the fission model was suspect because Earth and the Moon don't revolve fast enough. For coaccretion, the problem was the reverse: Earth and the Moon are spinning too fast. This means that the swarm of planetesimals around proto-Earth must also have been revolving quickly in the prograde (west to east) direction. But it was constantly being fed by planetesimals from the solar disk that would *not* have any preferred direction—they would plow into the swarm from all different directions, some west to east and some east to west. So it was not clear how the swarm could maintain its angular momentum, or even develop it in the first place.

Actually, there was a clever argument, originally conceived by Chamberlin, that suggested that planetesimals would be slightly more likely to strike Earth (or the protoplanetary swarm) in a west-to-east direction. Chamberlin reasoned that planetesimals that struck Earth at the *outermost* point of their orbits around the Sun would be moving slower than Earth; hence Earth would catch up on them from behind, and the impact would be west to east. And planetesimals that struck Earth at the *innermost* point of their orbits would be moving faster than Earth. They would catch up to Earth and, again, strike in a west-to-east direction. But computer simulations in the 1980s showed that the bias toward prograde impacts was very slight indeed, and not enough to provide the required angular momentum.

The second objection to coaccretion came from chemistry rather than physics. As Roche had been one of the first to realize, Earth has an iron core. According to current estimates, it makes up 32 percent of the mass of the planet. The Moon, on the other hand, has almost no core—roughly 2 to 4 percent of its mass. (It wasn't until the Lunar Prospector space mission in 1998 that scientists could say with any confidence that it has a core at all.) If Earth and the Moon both grew from material in the same feeding zone, why had almost all of the iron gone into Earth, leaving only rock for the Moon?

Ruskol proposed that the planetesimal swarm had acted as a filter. Iron meteorites punched through the swarm of planetesimals because they were more cohesive; rocky meteorites, on the other hand, tended to break up and leave more debris behind in orbit. Thus the Moon, which formed out of this leftover material, would have been mostly rock. But scientists in the West never warmed up to this idea. They didn't have a disproof, but it just seemed like an implausibility stacked on top of an implausibility. It was like a child trying to convince his or her mother that there is a monster in the closet. The mother looks in and says that she can't see it, and the child says that's because the monster is wearing an invisibility cloak. That line of reasoning might be very convincing for the child, but not for the parent.

Thus, on the eve of the Apollo missions and even for a good while afterward, there were three contenders for the explanation of the Moon's origin. Each of them had serious liabilities. This may explain why, when Mitroff polled the Apollo scientists in 1969, he found that by far the most common response to his questions was *indifference.* Half of the scientists he polled—twenty-one of forty-two—expressed no opinion or no interest as to which of the theories was most likely to be correct. He gave a lengthy "composite quote" that summed up the comments of these individuals: "You've got to realize that we've lived with some of these theories for so long that they don't mean much to us anymore. To a large extent these are matters of almost pure speculation. We've heard the same old people spin out the same old cobwebs and speculations for years without adding much to them. (Etc.)" So even while NASA was claiming in public that the Apollo missions would try to settle the question of the Moon's origin once and for all, the scientists who would be analyzing the Moon rocks were privately very doubtful. And for several years it looked as if they were right.

8

Renaissance and Controversy

The 1960s were the best of times and the worst of times for lunar science in America. The space program was like a rocket engine, energizing a discipline that had been almost dormant before 1950. The study of the Moon had a credibility and even an urgency that had never existed before. Men were going to be walking there by the end of the decade, after all, so it was essential for scientists to find out everything they could about it.

On the other hand, it was also a decade full of controversy. To begin with, scientists were always criticizing NASA for not paying enough attention to science, or for diverting resources from more worthy projects to an immense, money-sucking public-relations stunt, as some perceived Apollo to be. When they weren't battling NASA, scientists and engineers were battling each other, often along disciplinary lines. A passage from geologist Don Wilhelms's 1993 book *To a Rocky Moon* puts it well: "Sky scientists [i.e., astronomers] in particular regarded Apollo as a victory of the philistines over the forces of enlightenment, represented by themselves. On the other side, the Apollo engineers . . . had a world-shaking task to perform and did not appreciate the parochialism of scientists who emerged briefly from their ivory towers to view a world that was not crafted to their specifications." From the ivory tower side of the disciplinary divide, Harold Urey wrote this zinger to NASA's associate administrator, George Mueller, fewer than three months after the *Apollo 11* landing: "You have turned heavily to geologists. I know of some great, brilliant geologists, but mostly they are a second rate lot. . . . We all know that geology attracts the less brilliant type of scientists."

Even when the focus was strictly on science rather than professional qualifications, scientists could not agree on anything. There were three theories of the Moon's origin and two theories of craters. Some people believed in a "hot Moon" (i.e., one that had been molten in the past, and perhaps was still molten under the surface); others insisted on a "cold Moon." The lunar seas, or maria, were either lava flows or giant dust bowls. If they were lava flows, the magmas had either come from deep inside the Moon or just underneath the surface, or perhaps not from the Moon at all—perhaps they were liquefied remnants of meteorites that had formed the seas. Either the Moon had always been dry, or it had once held ice and even flowing water. Maybe tektites (strange glassy beads that are found in certain locations on Earth) came from the Moon, or maybe meteorites came from the Moon, or maybe tektites came from meteorites. . . . There seemed to be no end to the debates. But perhaps this is typical of any scientific renaissance.

Although it would seem natural to date the renaissance of lunar science to 1957, when the Russian satellite *Sputnik I* was launched, I would actually place it a little bit earlier. Public interest and scientific progress do not always move in phase. I would argue that the real turning point for scientists came in 1948, when an amateur astronomer and professional engineer named Ralph Baldwin published a book called *The Face of the Moon*, which brought the Moon back into scientific conversation and directly or indirectly influenced many of the leading scientists of the Apollo era.

Hiroshima on the Moon

Ralph Baldwin was a rarity in twentieth-century science: an outsider to the academic world who nevertheless managed to make academic scientists sit up and listen. At an early Moon meeting in 1962 a reporter asked him, "What school are you with?" When Baldwin explained he wasn't at any university, the reporter asked him, "How did they let you in?"

It might be more accurate to turn the anecdote around and say that it was Baldwin who "let in" the academics, or at least clued them in. (Don Wilhelms dedicated *To a Rocky Moon* "to the amazing Ralph Baldwin, who got so much so right so early.") Baldwin was, perhaps, what T. J. J. See might have been in a vastly different and more benign universe. Though he was an outsider like See, he was modest and soft-spoken, devoted most of his research to one topic, and did it well. That topic was lunar geography or selenography, and specifically the theory that the craters had been created by meteorite impacts.

Although Baldwin dedicated his whole career to the Moon, he certainly didn't start out as a fan of that body. When he was studying astronomy in college, he once said, "I didn't like the doggone thing. It got in the way, the light of the Moon. It bothered every kind of observation I wanted to make." This attitude was typical of most astronomers of that time. All of the action was in deep-sky astronomy, where scientists were making spectacular discoveries: galaxies beyond the Milky Way, the expansion of the universe, exotic kinds of stars. These were, for the most part, distant and faint, and could be photographed only when the Moon was not around.

In 1941 Baldwin took a job at the University of Chicago's Adler Planetarium and, purely out of curiosity, started examining the Moon photographs there. They included some beautiful eighteen-by-twenty-four-inch photographs that had been made at the Mount Wilson and Lick observatories in the early 1900s, the best images of the Moon's surface that existed until the U.S. space program began mapping the Moon in the 1960s.

In the high-resolution photographs, Baldwin discerned a curious pattern. There were numerous grooves on the surface of the Moon—long, straight valleys that all pointed like spokes of a wheel to one place: the center of Mare Imbrium, one of the "eyes" of the "man in the Moon." When he searched the literature, Baldwin found that only one other scientist had ever commented on these grooves, and he had described them incorrectly as parallel, rather than radiating from a common center. To Baldwin, the significance of the grooves was obvious. The mare must have been created by an immense explosion "of almost inconceivable violence," which had ejected mountain-size boulders that scoured out the valleys he saw in the photographs.

Baldwin presented his conclusions in a lecture at the University of Chicago, but it didn't exactly arouse an enthusiastic reception. For starters, he had to overcome the prejudice that there was no serious science to be done on the Moon: "The Moon was subject anathema," Baldwin has said. "I had changed my opinion but no one else had."

In just two months of reading, Baldwin had already learned more about the Mare Imbrium than anyone else knew, or wanted to know.

More specifically, his theory ran afoul of the prevailing dogma that all of the Moon's craters were created by volcanic activity. Volcanoes are the most obvious terrestrial analogue for these circular pockmarks on the Moon's surface, and they fit in well with the nineteenth-century assumption that Earth and the Moon had been molten bodies only a few dozen millions years ago. A hot, young Moon would still have plenty of energy to discharge through volcanic vents.

The most obvious problem with the volcanic theory was the scale of the Moon's craters, which are immense compared to any volcanic calderas on Earth. But geologists found other structures on Earth that could be analogues of the lunar craters. Some volcanoes, such as Vesuvius, are surrounded by a ring of lower hills, so that the ring and volcano together bear a vague resemblance to lunar craters with a rim and a central peak. (This analogy fails to explain, however, why the rims of lunar craters are always *higher* than the central peaks.) Or the lunar craters could be maars, which are circular depressions created by underground explosions of steam.

The idea that craters had been formed by meteorite impacts was almost as old as the volcanic theory, appearing for the first time in the scientific literature as early as 1829. However, very few people had taken it seriously. Until the early 1800s, meteorites had about as good a reputation among educated people as UFOs do today. "I would more easily believe that two Yankee professors would lie than that stones would fall from heaven," said President Thomas Jefferson in 1807, after two Yale University scientists claimed to recover 330 pounds of rocks from a reported fireball in Connecticut.

The Connecticut case, along with similar observed meteorite falls in France in 1790 and 1803, convinced many scientists that meteorites were real. However, meteorite *craters* were a different story, because there was not a single documented example of such a thing on Earth. How could the Moon be covered with the scars of giant meteorite impacts, while Earth had escaped without a scratch? One possibility, of course, was that Earth had many meteorite craters in the past, but they had all been eroded away. But until some concrete evidence was found, most scientists would opt for the more conventional volcanic theory.

One who did not, as we saw in the previous chapter, was T. J. J. See, who was already a pariah for plenty of other reasons. However, even a respectable advocate such as Grove Karl Gilbert, the president of the U.S. Geological Survey, was unable to arouse enthusiasm for impacts. Gilbert published an article in favor of the meteoritic origin of craters in 1893 that is today considered a classic. But his paper was virtually ignored in its time, and Baldwin didn't even find out about it until 1948, seven years after his initial study of the Mare Imbrium ejecta.

What Baldwin was asking scientists to believe in 1941 was even more radical: an impact had created not a mere crater, but a gargantuan structure that spans one-ninth of the Moon's circumference, a hole bigger than the entire state of Texas. It was too much to swallow. He was unable to get his paper published in a research journal and had to settle for the pages of *Popular Astronomy* instead.

After 1941, World War II interrupted Baldwin's budding career as an astronomer. Shortly after the war, his father asked him to take over the family business, the Oliver Machine Company, and after that, astronomy became for him only an avocation. But that did not diminish his passion for the Moon. In fact, in a bizarre way the war actually helped him gather evidence for the impact theory. At military bases all over the West, at bombing ranges and munitions testing facilities, there were fresh new craters that could be studied and compared to lunar craters. Later, as the Cold War wore on, craters produced by nuclear explosions would be added to the collection. Man-made craters such as "Jangle U" and "Teapot Ess" turned out to be perfect miniature replicas of craters on the Moon.

In 1948 Baldwin wrote *The Face of the Moon*, which documented in great detail how explosive craters are formed and what they look like. Most importantly, it did so in a quantitative, testable way. He worked out the exact relationship between the depth and the width of a crater and the energy in the explosion that created it. He showed that both man-made craters and lunar craters obeyed this relationship, even though they were at opposite ends of the spectrum in terms of size. In addition, he cleared up several misperceptions about impact craters, including some that had tripped up even Gilbert.

In the early 1890s Gilbert had visited Meteor Crater in Arizona (then known as Coon Mountain, and later as Barringer Crater) in the hope of understanding where this very Moon-like structure had come from. As in the case of the lunar craters, he considered two possibilities: it was a maar, or it was created by a meteorite impact. If it was the latter, he reasoned, then a large part of the meteorite should still be underground. Since the meteorites that make it to Earth usually have a large component of iron, this iron mass should create a local disturbance in Earth's magnetic field. Seeing no deflection of a compass needle, Gilbert concluded that there had been no meteorite. (Undaunted, the owner of the land, Daniel Barringer, would later form a company to mine the iron that he was sure lay beneath the crater—and he, too, came up empty.)

Gilbert and Barringer were misled by the fact that the only collisions we ever see in everyday life are low-speed impacts. Throwing a rock on the ground; dropping it from a tall building; even firing a bullet into the ground, as Gilbert did in some experiments—these are all *low-speed* collisions. The bullet may gouge a big hole, but the bullet remains basically intact; it doesn't explode. It transfers momentum to the soil, but not a lot of energy.

A high-speed impact, on the other hand, is much more like an explosion than a collision. The 63,000-ton stone that plowed into the Arizona plain fifty thousand years ago was moving at more than forty thousand miles per hour—fifty times faster than a speeding bullet. At that speed it contained ten times more energy per pound than a stick of dynamite. When the energy was released, the stone exploded. That is why no large body of iron was ever found underneath Meteor Crater, and only a few beach-ball-size fragments were found outside of it. The vast majority of the impactor had simply been vaporized.

Until 1945 it was impossible for humans to conceive of such an explosion. But now we can. Think of Hiroshima—only bigger. Like a nuclear bomb, the explosion would first produce a blinding spherical shock wave, then a mushroom cloud. The shock wave is what does most of the damage. It travels down into the ground as well as up into the air, subjecting the rock to immense pressure. It isn't really accurate to think of the meteorite "digging" a crater; instead, it compresses, fractures, pulverizes, and even melts the rock, and launches it in all directions. These secondary "bombs" create a lot of havoc in their own right. On the Moon (though not on Earth) it is easy to spot secondary craters that were created by material ejected from larger impact craters.

The planet takes a long time to rebound from the violence of the impact. Over time the crater gets shallower, because the weight that had been pressing down on the rock at the bottom of the crater has been removed. Physicists call this an "isostatic adjustment"; it has nothing to do with erosion, and thus it can happen just as well on the Moon as on Earth. This explains why many of the Moon's craters are very flat-bottomed.

Baldwin also explained why lunar craters are round, a fact that had puzzled Gilbert. Gilbert thought that a meteorite could create a circular crater only if it came in along a nearly vertical trajectory. An oblique or glancing impact would, he assumed, create a more elongated crater. Yet stretched-out craters are very rare on the Moon. The resolution of this paradox, Baldwin showed, is again that we are accustomed to impacts where the colliding bodies remain intact. A meteorite impact is first and foremost an *explosion,* and the shock wave produced by an explosion is symmetric. So the crater will be circular as well.

Baldwin's book was nothing less than a new way to think about craters, and it showed conclusively that craters on the Moon fit the profile of explosion craters on Earth. But there was still one missing piece of the argument. If the Moon had been so bombarded with meteorites that its whole surface was pocked with craters, so should Earth. But where were the meteorite craters on Earth? Fortunately, Baldwin's analysis told geologists what to look for. They should not waste their time looking for a projectile underneath the crater. Instead, they should search for the bizarre materials created by the shock wave from the impact, minerals that cannot ordinarily be created at Earth's surface because the pressures are not high enough. One such mineral is diamond, formed by squeezing graphite with a pressure of thirty thousand atmospheres. An even more convincing proof would be a mineral called coesite, which can be made by subjecting quartz to twenty thousand atmospheres of pressure. This had never been seen outside a laboratory, and (unlike diamond) could not be accounted for by any other known natural process.

In 1960 Eugene Shoemaker and Edward Chao announced the discovery of coesite at Meteor Crater, and the floodgates opened. Now geologists had a reliable way to identify impact craters. Shoemaker discovered another one later that year, the Rieskessel Basin in Germany, simply by taking a sample of the stone from a nearby cathedral. (The stone had been quarried from that basin.) By the end of the decade, forty-seven confirmed impact craters were known on Earth, in locations from Canada to South Africa. These were still very modest in size compared to lunar craters and to the granddaddy of them all, the Imbrian Basin. But there was no question anymore that extraordinarily large rocks can and do fall from the sky, rocks not merely the size of an oxcart or a house but also the size of a town or a city.

It was no accident that Shoemaker was the first geologist to positively identify impact craters on Earth. With any luck, he also would have explored them on the Moon. As he told his friend and biographer David Levy, he had lusted after the Moon ever since April 1948, when it suddenly struck him as he was going to breakfast one morning that scientists would really go to the Moon in his lifetime. "Why will we go to the Moon?" Shoemaker asked himself. "To explore it, of course! And who is the best person to do that? A geologist, of course!"

From then on, Shoemaker did everything he could to qualify himself to be that geologist. He read Baldwin's book. He crawled over nuclear-bomb craters and Meteor Crater. He founded the U.S. Geological Survey's astrogeology branch in 1960, when his colleagues at the Survey thought the whole thing was a crazy lark, that the United States would never send men to the Moon. He was one of the three lead scientists for the Ranger missions, the U.S.'s first unmanned spaceships to crash-land on the Moon. But in 1963, just two years before NASA chose its first scientist-astronauts, Shoemaker's dream was dashed. He was diagnosed with Addison's disease, a disease of the adrenal gland that (even though he was successfully treated for it) made him unable to qualify for a pilot's license. Instead of becoming the first geologist on the Moon, Shoemaker had to settle for choosing the first one. In the end it was Harrison Schmitt who lived his dream, walking on the Moon as part of the crew of *Apollo 17.*

By now, impact theorists have more or less won the debate about the Moon's craters. But in the 1960s, the debate was still in full swing. There were still plenty of "hot-Mooners," and plenty of room to maneuver between the extreme positions that all craters were volcanic and that they were all meteoric. The hot-Mooners could point to other evidence of volcanic activity, such as the dark maria, which most people (though not all) agreed were frozen lava flows. They could also show pictures of rilles, which might be lava tubes, and small "halo craters" with dark rings, which may have been active volcanoes at one time. More controversially, they also could cite "transient lunar phenomena," brief flare-ups that had been reported sporadically by telescopic observers as early as the 1600s and interpreted as volcanic eruptions. All of these phenomena suggested that volcanic activity was still going on, or at least had been going on in the geologically recent past.

Harold Urey and the "Crooked Moon"

Unlike Shoemaker, who was still at the beginning of his career when he read Baldwin's book, Harold Clayton Urey was already one of the most famous scientists in America. In 1934 he had received the Nobel Prize for his discovery of deuterium, a heavy form of hydrogen, which made him (at age forty-one) one of the youngest Nobel laureates ever. With his expertise in the separation of isotopes, he became a major contributor to the Manhattan Project during World War II, working on the gaseous diffusion method for enriching uranium. But like many atomic scientists, he was dismayed by the use of the atomic bomb against Japan. He lobbied against a law that placed the development of nuclear energy under military supervision, and joined organizations such as the Union of Concerned Scientists, which opposed the proliferation of nuclear weapons. In the years after the war he actively sought new areas of research, perhaps trying to distance himself from the nuclear-weapons establishment.

In 1949 Urey took *The Face of the Moon* with him on a train trip to Canada. According to a biographical sketch written by three of his friends, the book "started him on a love affair with that object which continued for the rest of his career." He educated himself rapidly about the Moon's surface, scrutinized the photographs that Baldwin had scrutinized, and was soon writing to Baldwin and grilling him about the geology of the seas and craters. He would regale his colleagues at the University of Chicago with his latest theories, even if they had no idea what he was talking about.

Urey had, by all accounts, a difficult personality. One newspaper reporter wrote that he had "a bulldog face and a bulldog mind," and the description seems well borne out by his correspondence. "I should be awfully glad if you would come to Chicago sometime to have a real good argument about this matter," he wrote to Baldwin in 1952. "It is so seldom that one finds anyone who has really tried to under-

Harold C. Urey (1893–1981), a Nobel laureate in chemistry for his discovery of deuterium, brought immense prestige and political clout to lunar science in the 1950s and early 1960s. His *cause célèbre* was the idea that the Moon was a "primitive object," possibly older than Earth, that had never melted and thus contained an unaltered chemical record of the newborn solar system. This idea did not stand up after the Apollo missions, but the chemical perspective he brought to planetary studies has remained very influential to this day. *Photograph courtesy of University of Chicago Library.*

stand the surface of the moon." The more mild-mannered Baldwin seems to have been a little taken aback by Urey's bellicose idea of a good time; a few years later one finds Urey mending fences: "You seem to be much more disturbed about the disagreement between us than I am."

But not all of Urey's conflicts were so easily smoothed over. He was forever breaking off relations with other people over slights real and imagined. His bitterest feud of all, which grew to legendary proportions among planetary scientists, was with Gerard Peter Kuiper, an astronomer at the University of Chicago who had been the biggest fish in the pond of planetary science before Urey came along. The two had started out in 1949 as friends, but Kuiper was a "hot-Mooner" and Urey was a "cold-Mooner." In 1954 and 1955 their scientific disagreement mushroomed into a very personal one; eventually the two stopped speaking to one another. Urey went so far as to accusing Kuiper of plagiarism and writing to the University of Chicago's chancellor to insist that the university press should not publish any more of Kuiper's books.

The feud created an especially awkward situation during the Ranger missions, on which they were two of the three lead scientists.

Shoemaker, the third scientist on the team, had to sit between them and act as a mediator. Both men left the University of Chicago rather abruptly: Urey in 1958, when he reached retirement age and jumped to the just-forming University of California at San Diego; Kuiper in 1960, when he pulled up stakes and founded the Lunar and Planetary Laboratory at the University of Arizona. This turned out to be a profoundly positive move, perhaps proving that every dark cloud has a silver lining. The Arizona institute prospered with NASA funding and turned out many of the top planetary scientists of the next generation, including Bill Hartmann, of whom much more will be said later, and Alan Binder, the mastermind of Lunar Prospector, NASA's only mission to the Moon since *Apollo 17.*

Urey brought the passion of an amateur and the outsized ego of a Nobel laureate to the study of the Moon, but he also brought one more thing that was critically needed: the perspective of a chemist. In Urey's solar system it was chemistry, as well as physics, that governed the accretion of planets. Urey believed that planets had started out as dirty snowballs, grains of dust that were glued together by ice. In his 1952 book *The Planets,* Urey speculated that the properties of water, ammonia, and methane ice had essentially determined where all the bodies in the solar system grew up and how large they came to be.

The Moon played a special role in Urey's system because he considered it the most primitive of all the inner planets—Mercury, Venus, Earth, and Mars being the other four. The reason for this conviction lay in the Moon's density. Earth is the densest body in the inner solar system, at a hefty 5.5 grams per cubic centimeter, and thus it must have a very substantial iron core. (Iron has a density of 7.8 at ordinary pressures; silicates, the minerals that predominate in Earth's mantle, have a density of about 3.3 to 3.4. Water has a density of 1 gram per cubic centimeter.) The Moon, on the other hand, is the lightest of the five inner planets, at 3.4 grams per cubic centimeter—about the same as ordinary rock. Thus, if the Moon had any iron core at all, it had to be a very small one.

If the story of the outer solar system was mostly about ice, the inner solar system was mostly about iron. Why did Earth have so much? Why was the Moon so anemic? Why had Earth's iron mostly collected in its core, as shown by seismographic studies? Urey devised an elaborate scenario that explained it all. The solar system, he believed, had begun as a cold dust cloud consisting of ices, silicates, and metal oxides. As mentioned above, the ices started to glue the other dust particles together into asteroid-size or even Moon-size bodies. Then the Sun "ignited" and started heating the nebula up. The inner solar system turned into a chemistry lab, where compounds such as water and ammonia boiled off (except where they had already been locked inside planetesimals), and the iron and silicates melted and separated from each other. The later a planet formed, the more its raw material had been distilled by this process and the higher its iron content would be. The Moon, with its small amount of iron, was therefore the most primitive planet and the only one that had never melted.

Urey also defended the cold-Moon theory with more direct physical observations. For example, in the 1950s it was believed that the Moon is ellipsoidal, as tidal theory predicts. This means it would have three different radii. The axis pointed toward Earth would be longest, and the north-to-south axis would be shortest. (The most recent data from *Lunar Prospector,* in the 1990s, do not bear this out.) From his observations, Baldwin had estimated the extent of this bulge as 2.2 kilometers, or more than 1 mile. Though this was not very much compared to the Moon's 1,737-kilometer radius, it was a larger bulge than expected from tidal theory. This suggested that the Moon had become "fossilized" in this shape at a very early stage in its history, when it was much closer to Earth and the tidal forces were larger. To maintain such a bulge over 4 billion intervening years, the Moon must have been very rigid the whole time and thus could never have melted.

Later, in the 1960s, Urey added more arguments for the cold-Moon hypothesis. When the unmanned Lunar Orbiter missions flew around the Moon in 1966 and 1967, mapping the potential Apollo landing sites, the mission controllers found it unexpectedly hard to keep them in a stable orbit. The reason was that the Moon contained unexplained concentrations of mass (which were named, rather unimaginatively, "mascons") centered in the maria. It was the most sensational discovery of a mission that had not really been expected to produce any discoveries. According to the principle of isostasy, which Baldwin had used to explain the flattening of craters, the maria should have rebounded to the point where there was no difference between the external gravitational fields above the lowlands and the highlands. Urey pointed to this lack of isostatic adjustment as another proof that the Moon was extremely rigid. It was more like a steel bar that does not bend under a lead weight, rather than a hammock that sags to accommodate it.

Also, both Soviet and American orbiting spacecraft showed that the Moon's "center of mass"—the fictitious point that acts as its gravitational center—is not at its geometric center, but is displaced by a kilometer or so toward Earth. The reason, it later turned out, is that the Moon's crust (which is made of lighter material) is thicker on the far side. But in any case, this seemed to Urey to be another proof that the Moon could never have been molten. If the Moon had melted completely, its mass would have been distributed uniformly, and there would be no difference between the geometric center and the center of mass.

"As to the moon," wrote Urey in his last letter to Baldwin, in 1972, "I approached it years ago without caring a little microscopic damn as to whether the moon was hot or cold. Then the moon had a tri-axial ellipsoidal shape. Since then, it has mascons, it has a center of mass displaced from the geometric center, it has gravitational anomalies in the big craters—Alphonsus etc., and the front side of the moon is depressed. . . . I refer to it as the crooked moon." By this time he already knew that the "crooked moon" had gotten the last laugh on him.

Urey, NASA, and Apollo

Urey's scientific legacy is complex, and it perhaps needs to be broken into two stages: short-term and long-term. Over the short term, many scientists were convinced that his cold-Moon theory was basically correct and that the Moon would provide a record of the earliest history of the solar system. Probably nobody agreed with every detail—lunar science was an amazingly contentious field—but his system became the closest thing to an orthodoxy from which other scientists would depart.

Urey's ideas found an especially eager reception at the fledgling National Aeronautics and Space Administration (NASA), which sprang into existence on October 1, 1958, thanks to an act of Congress, and which was desperately looking for a scientific rationale for its existence. (The political reason was obvious: the technological challenge posed by the Soviet launch of Sputnik the previous year.) Urey went to Washington, D.C., in December of that year to talk with NASA's deputy director, Homer Newell, and his main scientific adviser, Robert Jastrow. Out of that meeting came NASA's first attempt at a scientific mission statement: "We propose that a crash program be set up for the execution of a lunar soft landing in 1961. . . . The problem of the origin of the solar system is one of extreme importance to the origin of the Universe itself. It is one of the great problems with which the mind of man has been concerned since prehistoric times. Study of the moon's surface is intimately bound up to this problem. In fact, there is written plain to our eyes on the surface of the moon the history of the origin of the solar system." The words, Jastrow says, were mostly Urey's; Jastrow's contribution was to change "special program" to "crash program," for greater political impact. From that moment on, the Moon became NASA's primary objective. As Newell recalls, "The persuasiveness of the argument carried the day at each stage, within NASA, in the Administration, and finally in Congress." Surely the reputation of the arguer, a Nobel laureate, also made some difference.

For the sake of historical accuracy it is worth noting that Urey did *not* advocate sending men to the Moon. The statement above shows that he had in mind an *unmanned* mission at the earliest possible date, and a letter he sent the same month to then senator Lyndon Johnson makes this even clearer. In the letter Urey complained that the currently envisioned space program "will consist of the man-in-space project which is expensive, has little scientific interest, and as it is laid out does not have any continuing program beyond the immediate projects." The Mercury missions were already in the works, but no one saw them yet as stepping-stones to Gemini and Apollo.

The political debate was, of course, utterly transformed in May 1961, when President Kennedy made his famous commitment to send men to the Moon by the end of the decade. Historians have made much of the fact that this decision was shaped by politics, not science; indeed, it was Lyndon Johnson, now vice president, who identified the Moon mission as a goal America could reasonably beat the Soviet Union to. There also was a good deal of griping by scientists, throughout the decade, about the enormous commitment of resources (twenty-five billion dollars in the final analysis) to a project that did not have comparable scientific value. Urey agreed that the

Apollo program was too expensive to justify on science alone, but unlike other scientists, he accepted the political realities. However, he was a constant gadfly to NASA about other issues, such as the quality of their scientific advisers.

Perhaps it is time for a little bit of rerevisionist history. The Apollo program did have a legitimate scientific mission—the one Urey had identified way back in 1958. The purpose was to decipher the "Rosetta stone" of the solar system. But Urey was wrong to believe that an unmanned soft landing would accomplish that goal. Urey got his unmanned landing. In fact, there were five of them: the Surveyor missions between 1966 and 1968. But they did not resolve a thing: nobody changed his or her mind about the origin of the Moon or the solar system, least of all Urey. It took the manned Apollo missions to shake up the entrenched opinions of the academics. And Urey's cold-Moon theory would be among the first to go on the rubbish heap.

Nowadays, lunar scientists do not believe that the Moon has always been cold, for reasons explained in the next chapter. Nevertheless, Urey's legacy remains far more important than that of the other architects of failed lunar theories, such as Darwin and See, because Urey correctly identified the problems. In science, that is often more important than getting the right solutions. The large discrepancy between the amount of iron in Earth and the Moon remains one of the most important constraints for any theory of origins to explain. The timing and extent of core formation is another one. Besides asking the right questions, Urey also pioneered some of the right methods of answering them. He was one of the first scientists to use mass spectrometers to determine the chemical composition of rock, a dramatic improvement over the traditional mineralogical approach. His emphasis on meteorites as remnants of the early solar system also has stood the test of time. The oldest materials ever dated are the small "calcium-aluminum inclusions" found in certain meteorites, and they have established the currently accepted age (4.56 billion years) of the solar system.

Urey would be the first to admit that it was up to Mother Nature, not "Mr. Baldwin, Mr. Kuiper, or Mr. Urey," to provide the final answers to the questions of how the Moon and the solar system formed. He always bowed to real data, and after Apollo he very grudgingly admitted that his cold-Moon idea was flawed—the Moon had melted, at least on the surface. After 1972, Urey and his "bulldog mind" gradually had to let go of the scientific fray, as he began to suffer from the progressive effects of Parkinson's disease. He died in 1981.

Shoemaker, too, played a diminished role in the lunar debates after the Apollo missions. He had a very public separation from the NASA program in October 1969, when he announced his "retirement" to serve as chairman of geology at Caltech. Already at that time, when the news media were still all aglow over the success of *Apollo 11*, Shoemaker saw the handwriting on the wall and knew that NASA would never approve the advanced scientific missions he wanted as follow-ups to Apollo.

In the following years Shoemaker went to work on other projects, including the Voyager missions to the outer planets. (It turned out that his retirement from NASA was not permanent after all.) Together with his wife, Carolyn, and longtime colleague David Levy, he discovered Comet Shoemaker-Levy 9 and watched it spectacularly crash into Jupiter in 1994, leaving black clouds larger than Earth in its wake. This proved beyond a shadow of a doubt, if anyone doubted it anymore, that major impacts still happen in the solar system. Tragically, Shoemaker died only three years later, in an automobile accident in the Australian outback, while doing what he loved best: exploring terrestrial impact craters.

Though Urey and Shoemaker had done more than anyone else to ensure that the Apollo missions would be scientifically fertile, it would be up to the next generation to reap the harvest.

9

"A Little Science on the Moon"

On Earth it was August 2, 1971. On the Moon it was yet another sunny day with, as usual, no weather at all. (You can't have weather without water or an atmosphere.) Dave Scott and Jim Irwin, the commander and lunar module pilot, respectively, for the *Apollo 15* mission, were winding up the fourth visit by humans to the lunar surface, and they had saved a special event for last.

Standing in front of the garish orange lunar module, Scott held up two objects. "In my left hand I have a feather, in my right a hammer," he said. "And I guess one of the reasons we got here today was because of a gentleman named Galileo a long time ago, who made a rather significant discovery about falling objects in gravity fields. And we thought that where would be a better place to confirm his findings than on the Moon?"

The camera, controlled remotely from 238,000 miles away in Houston, zoomed in on the hammer and feather, and then zoomed out again just in time for the next act.

"I'll drop the two of them here and hopefully, they'll hit the ground at the same time." Scott later said that he was worried that the falcon feather might cling to his glove because of static electricity. Fortunately, it didn't. Though the hammer was about fifty times heavier than the feather, they dropped at an identical rate, the hammer sticking upright in the lunar soil for a moment before tipping awkwardly over.

"How about that!" Scott said, speaking over applause from Houston. "Mr. Galileo was correct." A few seconds later, he added, "Nothing like a little science on the Moon, I always say."

"Been saying it for years," radioed back Joe Allen, the physicist/ astronomer who was serving as capcom (capsule communicator) in Houston. Indeed, many scientists *had* been saying it for years, starting with Harold Urey and Gene Shoemaker and their tireless efforts to make sure that science stayed high on NASA's agenda. Nobody realized it yet, but the six Apollo landings would be the only opportunities to do "a little science on the Moon" for the rest of the twentieth century.

The Green, Green Glass of Hadley

In the course of three very busy days at the foot of Mount Hadley and the rim of Hadley Rille, Scott and Irwin had done way more than a little science. Like the three missions that had landed before them, they brought home a huge harvest of scientific information: photographs; descriptions of geological features; data transmitted from the instruments they placed on the Moon's surface; and most importantly, 168 pounds of rock and soil samples. This harvest would transform our understanding not only of the Moon's history but also of the whole solar system and our own planet.

In truth, Scott's hammer-and-feather experiment was not really typical of science on the Moon. To be sure, it was a fine tribute to Galileo, the man who conquered gravity intellectually long before NASA did it with rockets. And it was good public relations; it was a lot easier for the average person to comprehend than, say, the Suprathermal Ion Detector Experiment. But the outcome was a foregone conclusion. You don't have to go to the Moon to verify Galileo's theory; all you need is a sufficiently good vacuum chamber on Earth. George Adams, the inventor of one of the first air pumps, had done the same experiment two centuries earlier, using a guinea (a gold coin) and a feather. Adams's assistant had performed it for King George III of England, showing that British scientists, too, had a flair for public relations.

The more lasting payoff from the Apollo landings came from discoveries that no one had anticipated. For example, the previous day, August 1, while Scott and Irwin were exploring the steep slopes of the Hadley Delta, Irwin had noticed a large boulder on the hillside that looked green when viewed from a certain angle. "Can you imagine finding a green rock on the Moon?" Scott recounted later. "Think about that. We'd never had any green rocks in training. . . . Nobody has ever told you ever before, in any class that we could remember, [about anything green] other than olivine, and this clearly was not olivine—and, all of a sudden, you've got green! Man, that's something you go for regardless of how steep it is."

While controllers in Houston listened with a growing sense of alarm (the TV camera was turned off at the time, so they couldn't see what was happening), Scott parked the Lunar Rover on a fifteen-degree slope. This may not sound like much, but the slope was covered with loose powder. One wheel of the rover was dangling out in midvacuum, and Irwin had to hang onto the vehicle after they got out, to keep it from slipping downhill. Scott chipped off a one-pound fragment of the boulder, which became lunar sample number 15405, and dug up some green soil as well.

Scientists would discover later that the soil was full of tiny spheres of green glass, the frozen remnants of a "fire fountain" that erupted three and a half billion years ago, near the edge of the Imbrian Basin, in a spectacular sideshow to the formation of Mare Imbrium. The lava had cooled so suddenly as it vented into space that it rained back down in the form of glass beads. Scott was right that the rock was not olivine (a green mineral sometimes found in volcanic de-posits), but the glass beads that coated it had come from an olivine-rich source deep in the Moon's mantle. That was what made the find so remarkable: the depth and purity of the source. Thirty years later, geologists are still talking about it. To explain why, we have to back up to the first and most famous Moon landing.

First Impressions

In spite of its fame, *Apollo 11* was by far the least ambitious Moon landing from a scientific point of view. Mission planners for NASA picked the flattest, easiest landing site they could—the smooth, vol-canic plain of the Sea of Tranquility—and were prepared with back-up targets in case anything went wrong. Astronauts Neil Armstrong and Buzz Aldrin spent only two and a half hours walking on the Moon, barely an eighth of the time that Irwin and Scott would get. They collected 48 pounds of soil and rock samples to Irwin and Scott's 168. Their stripped-down experiment package contained only

A moment of high drama, for both science and lunar exploration.
Apollo 15 astronaut Dave Scott snapped this photograph, which
shows a green-colored boulder in the foreground, with his geologist's rake resting on top. In the background, astronaut Jim Irwin
struggles to keep the precariously balanced Lunar Rover from
sliding downhill. The samples Scott took from this boulder, and
the surrounding soil, turned out to be worth the trouble. They
contained beads of green glass that erupted from deep inside the
Moon 3.5 billion years ago, and provided crucial evidence about the
depth of the Moon's "magma ocean." *Photograph courtesy of NASA.*

three instruments: a seismometer, a solar wind collector, and a laser
reflector. By comparison, the *Apollo 15* astronauts had to set up eight
considerably more sophisticated experiments.

Nevertheless, first impressions are the most enduring, and that
was certainly the case with the first Apollo landing. Millions of people remember Neil Armstrong's ghostlike first step on the Moon,
while only diehard space junkies (including the author, who was a
twelve-year-old boy at the time) can recall Scott and Irwin's ramble
around the breathtaking vistas of Mount Hadley.

For scientists, *Apollo 11* may have been the most informative mission, simply because it was the first. It took only those two and a half hours of work by Armstrong and Aldrin, plus six intense months of study by a small army of scientists back on Earth, to answer fundamental questions about the Moon that had been debated for decades or centuries. How old is the Moon? What is it made of? What made the craters? What is the surface like? Is there any water? Any life? Later missions would modify and refine the answers, but *Apollo 11* would set the basic parameters.

Here is the Moon as *Apollo 11* revealed it (with a few added details from later missions):

How old is the Moon? The initial results were actually a bit confusing. Intact rocks from the Sea of Tranquility seemed to be about 3.5 billion years old, while the soil (or "regolith," a technical term first used by Shoemaker) was apparently 4.5 billion years old. It turned out later that the latter number was slightly in error, because *Apollo 11* landed in a region with higher than normal concentrations of the radioactive elements used to date the soil.

Nevertheless, some things were already clear. Nearly every Moon rock is older than nearly every Earth rock. A 3.5-billion-year-old rock is an extraordinary rarity on Earth, but this was the age of the youngest *Apollo 11* samples. Thus one hope of geologists was realized—the Moon gives us a window onto a period of time that has been virtually erased from Earth's geological record.

It also was apparent now that the Moon was not a mere youngster of one billion or two billion years, as some versions of the fission and capture theories had suggested. Likewise, the Sea of Tranquility was not a recent lava flow, so the "hot-Mooners" would have to look elsewhere for evidence that the Moon was still geologically active. On the other hand, *Apollo 11* was great news for Bill Hartmann, a former student of Kuiper who was just about the only scientist to correctly predict the age of Mare Tranquillitatis. (He did it by counting craters and estimating their rate of formation.) Even Shoemaker, who usually had good geological intuition, had believed that it was less than a billion years old.

What is the Moon made of? If *Apollo 11* caused some discomfort for the "hot-Mooners," it was an absolute Waterloo for Harold Urey, the leading "cold-Mooner." The samples brought back by Armstrong and Aldrin were made of basalt, a dark and porous rock formed by cooling lava. The chemistry of the rocks showed abundant signs of differentiation at high temperature, puncturing Urey's theory that they would be "Rosetta stones" revealing the primordial composition of the solar system. They were strikingly depleted of any element that boils at a low temperature. This absence of "volatile elements" turned out to be characteristic of all Moon rocks, and it would be an important clue about the Moon's formation.

From this mission and later ones, it became apparent that the Moon is much simpler, mineralogically, than Earth. Ninety-eight percent of the crystalline material on the surface is made of only four minerals. Later in this chapter it will be helpful to know what the four building blocks of the Moon are:

1. *Plagioclase feldspar.* This is a calcium aluminum silicate ($CaAlSi_2O_8$), in which the calcium may be replaced by sodium. On the Moon, sodium, a volatile element, is uncommon, so most of the plagioclase is of the calcium form, called *anorthite.* A rock that is more than 90 percent anorthite is called an *anorthosite.* (Confused yet?) Such rocks would become one of the most sought-after commodities of the Apollo missions, for reasons explained below.

2. *Olivine.* This green mineral is a magnesium-iron silicate, $(Mg,Fe)_2SiO_4$, in which the iron and magnesium are interchangeable. Naturally occurring olivine contains mostly magnesium, but the iron is responsible for the green color.

3. *Pyroxene.* An iron or magnesium silicate with a different formula ($FeSiO_3$) and that also can contain calcium ($CaFeSi_2O_6$).

4. *Ilmenite.* A dark, heavy, iron-titanium oxide ($FeTiO_3$). For unknown reasons, some mare basalts (such as *Apollo 11*'s) have a lot of ilmenite in them; others (such as *Apollo 12*'s) do not.

The key difference among these four minerals is that plagioclase contains no iron and the other three do. This makes plagioclase significantly lighter. The density of pure anorthite is 2.76 times the density of water, while the mare basalts are 3.35 times denser than water.

What made the craters? Apollo 11 found lots of evidence of meteorite impacts and no evidence of recent volcanic activity. Aside from the mare basalts, which are younger and less beaten up, most of the Apollo rocks were what geologists call *breccias.* These are formed when the shock wave from a high-speed impact fuses together unrelated rocks. On Earth they are found only near meteorite craters; on the Moon they are everywhere. The astronauts also noticed that many of the craters had small, glassy pools in the bottom—another result of the rapid heating and cooling from meteorite impacts.

All of this evidence showed that virtually nothing on the Moon's surface had gone untouched by the ages-long bombardment from space. Impact craters are the rule, and volcanic craters—if any exist at all—are rare exceptions.

What is the surface like? Thomas Gold, a lunar scientist with a penchant for unconventional theories, had suggested in the early 1960s that the maria were covered by a deep layer of dust that would swallow up any spacecraft that landed on them. This concern was put to rest by the unmanned Surveyor missions that landed successfully on the Moon in 1966 and 1967. When the *Apollo 11* lunar module landed on the Sea of Tranquility, one of its footpads burrowed about an inch into the soil. So much for the "Gold dust."

The lunar surface is indeed covered by a thick layer of pulverized rock fragments (the seismometer on *Apollo 14* would estimate the depth of this layer at nearly 30 feet), but below the first couple of feet the powder has been compacted to cementlike toughness. This material is quite different from Earth's soil: it contains no organic material or water, but lots of crushed rock, glass beads, and even grains of pure metallic iron. (You would never find pure iron in Earth's soil because it would rust.) That is why lunar scientists use the word "regolith" (literally, "blanket of rock") instead of "soil." Every scoop of the regolith is a pastiche of the whole lunar surface; even the samples collected by Armstrong and Aldrin in the low-lying Mare Tranquillitatis contained grains of rock from the highlands. Those grains were another clue to the Moon's early history.

Is there any water on the Moon? The first impression from *Apollo 11* was an emphatic *no*. The Moon is drier than the harshest desert on Earth. There are no bodies of water, no ice crystals, no water vapor, no sedimentary rocks, and no hydrous minerals—that is, minerals such as mica that appear dry to the naked eye but contain water in their chemical structure. This also means that it is unlikely anyone will ever find any lodes of silver or gold or other valuable metals on the Moon; on Earth, concentrated ore deposits are formed by the action of water.

In the 1990s, two unmanned Moon orbiters would discover an important exception to the Moon's nearly complete absence of water.

The U.S. Department of Defense satellite *Clementine,* which orbited the Moon in 1994, picked up some ambiguous radar signals that looked as if they might be caused by ice near the South Pole, and the *Lunar Prospector* mission confirmed this discovery. Scientists still do not know for sure how extensive the deposits are. It is believed that this water has been delivered to the Moon by comets, and was prevented from evaporating and escaping the Moon's gravity by being deposited in permanently shadowed craters near the pole.

Is there any life? With no water it is hard to see how there could be life on the Moon. But just to make sure, biologists scrutinized the Moon samples for any sign of biological activity. Mice and cockroaches were injected with Moon dust to see if they would be infected by germs from the Moon. The astronauts from *Apollo 11* were kept in quarantine for three weeks and the rock samples for even longer, lest they expose the world to some unknown pathogen. Nothing living was ever found, and the precautions were eventually dropped after the *Apollo 14* mission.

In fictional voyages to the Moon, from Kepler's *Somnium* through H. G. Wells's *The First Men in the Moon,* rule number one had always been that the protagonists encountered some form of life. Even in the first century A.D., Plutarch had ended his *De Facie* with a story where the Moon was inhabited by ghosts. Apparently it was just too hard for the human intellect to grasp a place that was utterly devoid of life. Or perhaps the authors couldn't think of an interesting plot that didn't involve little green men or bug-eyed monsters. After July 1969, the Moon of fantasy was dead. But what remained was an unexplored world full of tantalizing geology—a planet whose history is older and simpler than Earth's, but just complicated enough to keep scientists guessing.

Eighty-four Grains of Sand

"To see a world in a grain of sand/And a heaven in a wild flower,/ Hold infinity in the palm of your hand/And eternity in an hour" wrote the English romantic poet William Blake. If there is one ability that both scientists and poets have in common, it is to grasp the cosmic import behind the smallest of things. Few have lived up to Blake's words more completely than John Wood, a meteoriticist at the Harvard-Smithsonian Center for Astrophyics.

When analyzing the lunar soil samples from *Apollo 11,* Wood noticed some tiny white flecks mixed with the charcoal-colored base.

He painstakingly counted each one of them; there were 84 white grains in a scoop of 1,676 particles that Armstrong and Aldrin had dug out of the Moon. These particles literally were sand; the largest ones were about 2 millimeters wide, the size of a largish grain of sand that you would find at the beach. Some of them were bits of rock; others were bits of glass, melted by impacts. Their chemical composition was the same as anorthosite.

The anorthosite didn't seem to "belong" in the low-lying maria, so Wood theorized that it had been blasted out of the lunar highlands by meteorite impacts. This made sense; as anyone looking at the Moon can tell, the highlands are lighter in color than the maria. Not only that, but the *Surveyor 7* unmanned mission, which landed at the young crater Tycho in the lunar highlands, had found soil there that chemically matched the 84 grains of sand in Wood's laboratory. The similarity of samples taken hundreds of miles apart made Wood wonder whether, perhaps, the whole lunar crust was anorthositic.

If so, there would be huge consequences. Geologists understand very well where anorthosite comes from on Earth. It is found where a pool of magma has sat around long enough for the light materials to separate from the heavy ones, and it is found only in local deposits. If all of the lunar highlands were made of anorthosite, it would mean that the whole crust of the Moon must have floated atop an ocean of magma.

It was a very daring theory—a planetwide ocean from eighty-four grains of sand! According to Wood, "opinions in the [lunar science] community ranged from intrigued to incredulous." But Wood wasn't the only one to come up with this idea; independently, Joseph Smith of the University of Chicago had suggested the same thing. In any case, it was an easy theory to test. Beginning with *Apollo 15*, the astronauts were trained to look for the white "genesis rocks" of anorthosite that might remain from the original crust. Scott and Irwin found one, even though their landing site in a small "bay" at the edge of the lunar Apennine Mountains was not, strictly speaking, in the highlands. This find turned the tide of scientific opinion and made the magma ocean seem not only credible but also likely.

One remaining criticism was that the global layer of anorthosite need not have formed all at once. Perhaps there were separate magma seas at various times, rather than an ocean that completely engulfed the Moon. But that is where the *Apollo 15* green glass comes in. According to a new (as of 2002) theory of John Longhi, a geologist at Columbia University, it could only have been formed at the bottom of a very deep, worldwide magma ocean.

Here is the reasoning: Whenever light minerals float to the top of a magma chamber, the heavy minerals also have to concentrate at the bottom. So if there was a crust of light plagioclase at the top of the magma ocean, there had to be a layer of heavy olivine at the bottom. The astronauts could not go looking for samples from the *bottom* of the magma ocean, of course, because they did not have the equipment to drive a bore hole miles into the lunar crust. However, the samples could come to them—and that is why the green glass was so important. It was a special-delivery package from deep in the Moon's mantle, and it showed that the concentration of olivine at that depth was amazingly high by terrestrial standards. The purity of the olivine is directly related to the depth of the magma ocean, which Longhi's calculations showed was more than six hundred miles. By comparison, the Pacific Ocean is a mere seven miles deep at its deepest point. If this is correct, it virtually rules out the idea that the Moon's anorthosite could have formed by separate local melting events. A reservoir of that depth (more than half the distance from the surface to the center of the Moon!) would have to be a worldwide ocean.

Another recent proof of the magma ocean hypothesis has come from *Lunar Prospector*, a NASA satellite that orbited the Moon in 1998 and 1999. This mission mapped the chemical composition of the entire Moon's surface and showed that all of the Moon's highlands, which cover 85 percent of the surface and almost the entire back side, are rich in aluminum—the signature of plagioclase. Because this was the first global-scale evidence for the global ocean, many planetary scientists feel that it has settled the debate.

The magma ocean, whose existence no one had dreamed of before *Apollo 11*, posed a difficult physical challenge for any theory of the Moon's origin: Where had enough heat come from to melt the whole surface, perhaps to a depth of six hundred miles? For the fission theory, this was no problem; Darwin had always assumed that Earth and the Moon were completely molten to begin with. For the capture theory, it was nearly a kiss of death. The whole point of the capture theory, as far as Urey and many other supporters were concerned,

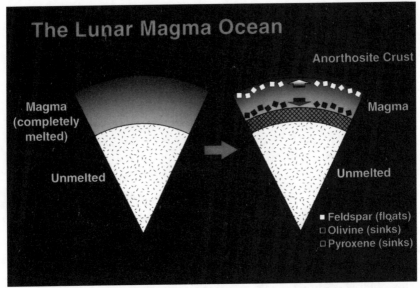

The magma ocean theory. The young Moon was covered by an ocean of magma that may have been six hundred miles deep. As the magma cooled, lighter anorthosite floated to the top, while the heavier iron-containing minerals olivine and pyroxene sank. *Apollo 15* brought back samples of both types. The widely publicized "genesis rock" is a sample of the anorthosite crust, while the green glass beads came from the deeper olivine layer. *Artwork by Brooks G. Bays Jr., University of Hawaii, courtesy of Planetary Science Research Discoveries.*

was that it enabled the Moon to stay cold throughout its lifetime. As for the coaccretion theory, the magma ocean was a serious but not necessarily fatal blow. If the Moon had accreted on the stately time scale proposed by Safronov for Earth, a hundred million years or so, then it would have been too cold. If it had formed in a great hurry, on the scale of years to centuries, then the intense bombardment of planetesimals could have heated the growing Moon enough to create a magma ocean. But what could have provoked the Moon to assemble itself in such a brief instant of geological time?

The "Late Heavy Bombardment"

Although the press made a fuss over "genesis rocks," the discovery of the magma ocean was not the only big conceptual breakthrough that came out of the Apollo missions. Another strange and very unexpected finding emerged from geochronologists' attempts to establish a timeline for the Moon's history.

The greatest challenge in lunar chronology was to move from dating *rocks* to dating *events*. Rocks can be dated in a laboratory by analyzing their radioactive isotopes, but events require careful geological interpretation, so that a particular sample can be tied to a particular crater or lava flow. For example, Gene Cernan and Harrison Schmitt of *Apollo 17* found many rocks that were believed to lie in a ray of debris from Tycho, the Moon's brightest crater, fourteen hundred miles away. The rocks had a uniform age of 109 million years, so that is most likely when the Tycho meteorite hit the Moon. If the astronauts had not been there to describe where they took the rock samples from, scientists might have erroneously attributed this age to a nearby crater instead of to the distant Tycho. (Imagine tracing a rock found in Iowa to a crater in Maryland.)

By 1974, through this combination of lab work and geological reasoning, lunar scientists had more or less agreed on how old the major basins on the near side of the Moon are. Strangely, they all seem to have about the same age: Mare Nectaris comes in at 3.92 billion years, Mare Serenitatis at 3.85 billion, Mare Imbrium at 3.83 billion. All seven basins formed in the brief interval between 4 billion and 3.8 billion years ago. This coincidence led Gerald Wasserburg of Caltech, the leading expert on radioactive dating, to propose that a violent spasm of meteorites had bombarded the inner solar system during that period.

Wasserburg's "late heavy bombardment" stirred up a controversy that has refused to go away. In 1975 Bill Hartmann argued that the cluster of 3.8-billion-year dates represented the tail end of a bombardment that had been going on since the Moon's formation. Larry Haskin of Washington University in St. Louis has claimed that the purported ages of the seven seas really date only one event, the formation of the Imbrian Basin. According to Haskin, the Apollo missions never landed far enough from Mare Imbrium to collect samples that were untainted by that enormous cataclysm. The *Lunar Prospector* chemical maps from the 1990s bear this out: All of the Apollo missions landed within a province of elevated radioactivity that is centered in Mare Imbrium and the adjacent Oceanus Procellarum. It is possible that we really haven't explored the whole Moon yet, we have only explored the Imbrium-Procellarum terrain.

Among the supporters of Wasserburg's theory, Graham Ryder of the Planetary Science Institute pointed out in 1979 that all "impact melts" collected from the Moon seem to have a similar upper age limit of four billion years. Any rock samples older than that do not seem to have been battered hard enough to melt. And in 2000, Barbara Cohen of the University of Tennessee analyzed several of the twenty-five or so known meteorites that have come from the Moon and found that their impact melts, too, go up to only four billion years of age. This would reduce the power of Haskin's argument, because the meteorites are probably a random sample from all over the Moon, including the back side; only a few at most are contaminated by the Imbrian event.

All in all, the jury is still out on Wasserburg's idea, and it will probably remain that way until humans return to the Moon and collect samples from all over the globe. It would be especially desirable to find out the age of the Moon's largest and deepest basin (actually, the deepest basin in the whole solar system): the South Pole–Aitken Basin on the far side of the Moon, whose existence was barely suspected at the time of the Apollo missions.

If it really did happen, the late heavy bombardment has profound importance for the history of the solar system. First of all, the end of the bombardment 3.8 billion years ago would set the time when life could finally appear on Earth. (Nothing could have survived impacts as large as the ones that excavated Imbrium or South Pole–Aitken. If the Moon had two impacts of this scale, Earth must have had several, because it is a larger target than the Moon and its effective size is enhanced by gravitational focusing.) Apparently life took advantage of the opportunity very quickly, because there are signs of biological activity on Earth as long as 3.5 billion years ago. It also seems hardly coincidental that the most primitive life forms on Earth seem to be the thermophilic (heat-loving) bacteria that crowd around deep-sea volcanic vents. Could they have been the first organisms that could survive on a planet that was cooling down from the late heavy bombardment?

What triggered the bombardment is still a mystery, but recent speculations have become more and more dramatic. In 2001 Harold Levison, Martin Duncan, and Luke Dones proposed that the bombardment had been triggered by the late formation of the giant planets Uranus and Neptune. Safronov's accretion models, as you might recall, had led to the faulty conclusion that Uranus and Neptune do not exist yet. Obviously that was too extreme, but it seems likely that they *did* take longer to form than the other planets. So Levison's group built a computer model wherein they put the other planets into their present orbits and then "switched on" Uranus and Neptune. Sure enough, this disturbed the asteroid belt enough to deluge the inner solar system with asteroids. The model worked even better if they made Uranus and Neptune form *inside* Saturn's orbit. For reasons not understood yet, the two planets seem to "want" to migrate outward, and again this process greatly destabilizes the asteroid belt.

Not to be outdone, John Chambers of NASA Ames Research Center proposed yet another migrating-planet theory at the 2002 Lunar and Planetary Science Conference. In Chambers's computer simulations, it would have been possible for a Planet V, located between the current orbit of Mars and the asteroid belt, to survive for six hundred million years, from the formation of the solar system until the initiation of the late heavy bombardment. It does require a somewhat careful choice of parameters: The simulations work best if Planet V was half the size of Mars, orbited at 1.9 times the radius of Earth's orbit (compared to 1.5 for Mars), and had an orbit that was slightly tilted with respect to the ecliptic plane. After six hundred million years, resonances with Mars's orbit would cause Planet V's orbit to decay, the perihelion moving inward toward the Sun and the aphelion moving outward, until it either collided with Mars (in about half the simulations) or plunged into the Sun (in the other half). Either way, it would sweep up a lot of asteroids on its final rampage through the solar system, and these would produce the late heavy bombardment.

Mark Twain, ever the curmudgeon, once wrote: "There is something fascinating about science. One gets such wholesale returns of conjecture out of such a trifling investment of fact." No doubt Twain would have found these two scenarios—conjectures built on top of conjectures—to be a case in point. Even Dones admits that their models would have been laughed out of the meeting fifteen years ago. They are taken seriously now, I think, for two reasons. First, scientists are growing more and more comfortable with an early solar system that appeared different from the system of today. Perhaps the discovery in recent years of very different planetary systems around other stars has encouraged this trend. Second, scientists increasingly accept (and rely on) computer models when classical math, such as Safronov used, is not good enough.

Curiously, the late heavy bombardment has not played much of a role in the debate over the Moon's origin. Superficially, it might seem to favor the simultaneous development, or coaccretion, model, because in that theory there is bound to be a lot of debris still in Earth orbit after the Moon forms. However, it is difficult to keep the debris in orbit for six hundred million years and then have it suddenly start pounding the Moon. In fact, the coaccretion model would be more consistent with a gradual tapering off of the collision frequency over time, as in Hartmann's rebuttal to Wasserburg's theory. If there was a lunar cataclysm, apparently it was triggered by events outside the Earth-Moon system.

Lunar Chemistry

Along with the "macrothinking" that went into the magma ocean and late heavy bombardment theories, there was also a lot of "microanalysis" to be done on the Moon rocks. The chemical mix of the Moon is bewildering, but cosmochemists simplify things by placing elements into certain broad categories: elements with a high boiling point (refractories) or a low one (volatiles); elements that like to alloy with iron (siderophiles) or that like to enter into rock (lithophiles). If the Moon had been undifferentiated (i.e., cold), as Urey hoped, all of these elements would have occurred in quantities similar to meteorites, which are the most primitive objects we know of in the solar system. Alas, such was not the case. The Moon has a lower concentration of volatile elements than either meteorites or Earth. The Moon's siderophile budget is likewise lower than that of meteorites but similar to Earth's. This last fact posed another puzzle. We don't see many siderophile elements in Earth rocks because they followed the iron when it separated and formed Earth's core. But the Moon's core is very small compared to Earth's: Where, then, did the Moon's siderophile elements go?

Of the three theories for the Moon's origin, Darwin's fission hypothesis seemed to pass the chemistry test most easily. The lack of volatiles again suggested a high-temperature origin. Moreover, if the Moon had separated from Earth after the core had formed, then it would essentially be a chunk of Earth's mantle, already depleted in iron-loving elements. In addition to these two constraints, one more chemical test also pointed to the conclusion that the Moon is a chip off the old block. In 1975 Robert Clayton and Toshiko Mayeda measured the amount of two heavy-oxygen isotopes (oxygen 17 and oxygen 18) relative to the "normal" oxygen 16 in a variety of Moon rocks. Oxygen isotopes have proved to be very sensitive tools for determining where in the solar system a rock came from. Earth rocks, meteorites from Mars, meteorites from the inner asteroid belt, and meteorites from the outer asteroid belt all have different oxygen isotope ratios. Clayton and Mayeda found that the isotope ratios for Moon rocks are indistinguishable from Earth rocks. In 2001 a Swiss team led by Uwe Wiechert and Alex Halliday repeated the analysis with instruments that were a generation more sensitive—and the conclusion remained exactly the same.

Will the Real Moon Please Stand Up?

One of the main scientific goals of the Apollo missions had been to determine, once and for all, where the Moon came from. The television commentators always said so, and NASA's own lunar exploration plan from 1965 said so, too: "The primary objective . . . is to define the nature, origin, and history of the moon as the initial step in the comparative study of the planets." As the dust settled on the last footsteps of the astronauts and as the Sun set over the last of their Moon bases, the nature and history of the Moon were much better understood than before, but the origin was still a mess. In all fairness to the scientists, it was a tremendously complicated problem—more complicated than finding a date, analyzing a chemical composition, or explaining a geological formation.

Of the three major origin theories, only the fission hypothesis had not been weakened by the Apollo missions. Unfortunately, it was the weakest going in, and many scientists flatly considered it to be impossible. The old problem of how Earth could have been spinning fast enough to throw off the Moon was still unsolved; moreover, the fission theory required Earth to begin in a molten state, and there was absolutely no evidence that it ever had been.

The capture hypothesis had taken much more serious hits. The discovery of the magma ocean was damaging, and the similarity between Earth's and the Moon's oxygen isotope ratios was a crusher. If the Moon had wandered in from somewhere else, such as the asteroid belt, the oxygen isotopes should have announced that fact loud and clear.

The coaccretion theory also was listing badly. The magma ocean meant that the Moon had to collect itself on an unaccountably rapid time scale. The problem of the iron-poor Moon, which had been recognized before Apollo, was now compounded by the additional mysteries of a volatile-poor and siderophile-poor Moon. Supporters of the theory continued to tinker with it to adapt to these facts, but their efforts appeared increasingly like "just-so stories."

All in all, the situation was reminiscent of the old TV game show *To Tell the Truth*. At the end of the episode, the real McCoy is supposed to stand up while the two impostors stay seated. But this time, none of the candidates had stood up. All three were looking like impostors! It was time for the *real* origin of the Moon to sneak in from offstage, where it had been waiting, unnoticed by anyone in the lunar science community, since 1946.

10

When Worlds Collide

At midcentury, Reginald Aldworth Daly was one of the deans of American science. In 1950, when *Scientific American* ran a series of retrospectives on the first half of the twentieth century in all the sciences, it tapped Daly to write the geology article. Born in Ontario in 1871, Daly spent most of his professional career at Harvard—first as a graduate student, then as an instructor, and finally for thirty years as a professor of geology. He was best known as a specialist on igneous rocks, but after he retired in 1942, he began to think about the Moon as well.

A seasoned geologist, Daly questioned the then conventional wisdom that the Moon's craters were volcanic, for basically the same reasons that Ralph Baldwin did. (The two men had no contact with each other.) The craters were too round, their rims were too high, and they were much too large to be volcanic calderas on the terrestrial model. Thus Daly, like Baldwin and Gilbert and See before him, concluded that they must have been formed by impacts. But he asked one more question that the other crater theorists had not specifically addressed: Where had the rock fragments come from to make all those craters?

Daly speculated that the rocks had been torn loose from Earth in a violent collision. In 1946 he published a fifteen-page paper in *Proceedings of the American Philosophical Society,* in which he wrote prophetically, "it might be worth while to study the question whether the main part of the Moon's substance represents a planetoid which, after striking the Earth with a glancing, damaging blow, was captured." Note that Daly, who called this "hypothesis No. 3" in his paper, phrased it as a variation on the gentler and more acceptable capture theory. However, he went on to make the violence of the impact clear: "Fragments [of the Earth] were torn off by the visitor, along with others ejected because of an explosion resulting from the collision. . . . It does not seem wild to assume a temperature of at least 100,000 degrees centigrade" in the vicinity of the impact.

Daly's paper marks the debut in the scientific literature of what has now become known in scientific circles as the "giant impact hypothesis" or, more informally, the "Big Whack." (I will call it the "Big Splat" because I think that is a more accurate and colorful description. However, I will also refer to it, using accepted terminology, as the "giant impact.") Unfortunately, despite the well-established reputation of its author, the Big Splat was a Big Dud. The reasons why make an interesting story within the story.

For two years before the paper appeared, Daly corresponded with Henry Russell, a famous Princeton astronomer who ultimately acted as one of the paper's reviewers. Russell clearly was not impressed, and toward the end Daly's letters began to sound very apologetic. "Perhaps the best thing to do is burn the manuscript," he wrote in December 1945. A few days later, still with a large dose of humility, he wrote, "I make three guesses [to explain the Moon's origin], chiefly to ask you master minds to produce the fourth and objectively valid guess. We geologists are or should be on tiptoe for the fourth! If there are no wickets put up there can be, will be, no cricket game."

In spite of Daly's plea, there was no cricket game. No astronomers or physicists wrote either to agree with or to rebut his theory. His article simply disappeared into oblivion. When the giant impact hypothesis was revived in 1974, none of the people who proposed it knew of Daly's prior paper.

It was a bad time to propose such a theory for many reasons. In the 1940s, the Moon was still totally déclassé among astronomers, as Baldwin also had found out. Most scientists still believed that the Moon's craters were volcanic in origin; from this point of view, there was no reason to believe that the Moon had been bombarded by Earth debris, and therefore no reason to suppose that a giant collision had put the debris there. Daly had gone at least two steps farther down the speculative road than his readers (if there were any) were prepared to go. It would take more than a fifteen-page paper to start changing scientists' minds about the Moon; only Baldwin's exhaustively documented book in 1948 would get their attention.

Second, if studying the Moon was déclassé, the idea of worlds colliding, or cosmic catastrophes in general, was completely beyond the pale. The dominant ideology in geology ever since the days of Charles Lyell, in about the 1830s, was that Earth was shaped by uniform processes, not by cataclysmic events. Nothing demonstrates this mind-set more clearly than a furor that erupted just four years after Daly's paper appeared.

The Velikovsky Affair

In 1950, Immanuel Velikovsky, a Russian-born psychologist, published a sensational best-seller called *Worlds in Collision,* which argued that the miracles described in the Bible and other ancient texts— events such as the "long day of Joshua," when the Sun and the Moon allegedly stood still in the sky—were in fact accounts of real events with astronomical causes. Velikovsky also discovered that many ancient civilizations had described the heavenly bodies differently from the way they appear today. Venus was represented as a "torch star," a "bearded star," or a "long-haired star." Although the identification of deities with objects in the sky is notoriously malleable, Velikovsky took these descriptions seriously. He argued that Venus must have appeared as a comet, rather than a planet. Since mythology said that Diana (often identified with Venus) had been born from the brow of Jupiter, that "birth" must correlate to a real event that ancient people had witnessed.

Tying all of these fragments together into a neat package, Velikovsky theorized that Venus had detached itself from Jupiter around thirty-five hundred years ago and made two close passes to Earth as a comet before settling down into its present-day orbit. One of the passes was close enough to stop Earth's rotation temporarily, explaining why the Sun appeared to stand still in the sky for Joshua. The comet's tail, he argued, was the source of the plague that descended on Egypt and the "manna from heaven" that nourished the Israelites during their wanderings in the desert. Venus's near approach also could have caused the floods and calamities described in many myths around the world.

Scientists could not ignore Velikovsky, as they had ignored Daly, because *Worlds in Collision* was shooting to the top of the best-seller lists. Harlow Shapley, director of the Harvard Observatory and author of a leading astronomy textbook published by Macmillan (which also published *Worlds in Collision*), denounced Velikovsky's book as "nonsense and rubbish." He threatened to withdraw the rights to his textbook from Macmillan if the company continued to print *Worlds in Collision*. Other scientists threatened to follow suit. Faced with the possibility of a boycott by its leading academic authors, Macmillan backed down and—in a move that would be inconceivable in today's publishing world—sold the rights to Velikovsky's book to Doubleday.

But that wasn't the end of the brouhaha. Shapley had never intended to suppress Velikovsky's work entirely, only to remove the stamp of legitimacy it might have gained from being published by Macmillan. But to the public, it looked like a crude attempt at censorship. If anything, the episode made the book even more popular, and gave Velikovsky permanent legitimacy as a counterculture hero, a sort of modern-day Galileo.

Surprisingly, Velikovsky did find one bastion of support in the scientific community during the furor: Albert Einstein. The two had become acquainted in the 1920s, when Velikovsky was the founder and Einstein an editor of a journal that promoted Jewish scholarship. Later, in 1946, Velikovsky sent a draft of *Worlds in Collision* to Einstein, to which the famous scientist replied with remarkable tact and prescience. Einstein warned Velikovsky that nobody would take the scenario involving the planet Venus seriously, and it would detract from his more persuasive arguments about the reality of ancient catastrophes. He advised Velikovsky to remove the Venus theory from his book: "If you cannot decide on this, then . . . it may be difficult finding a sensible publisher who would take the risk of such a heavy fiasco upon himself."

When Velikovsky and Einstein met again in 1952, while the fiasco Einstein had predicted was in full bloom, Einstein was at first angry, thinking that Velikovsky had set out to dupe the public. But the two men quickly warmed to each other, as Velikovsky described years later in a memoir called "Before the Day Breaks." The title alludes to a biblical story of a wrestling match between Jacob and an angel, which Velikovsky sent to Einstein as the latter's health was failing:

And [the angel] said, "Let me go, for the day breaketh."

And [Jacob] said, "I will not let thee go, except thou bless me."

Velikovsky saw himself as Jacob, and Einstein as the angel whose time on Earth was nearly over.

Though Einstein did bless Velikovsky with a sympathetic ear, there is no indication that he ever blessed his theories. For three years, until Einstein's death in 1955, the two did polite intellectual battle, mostly by letter. Einstein's position remained, as he succinctly stated in a 1954 letter, "Catastrophes *yes*, Venus *no*." Even this, Velikovsky replied, was more than other scientists would concede: "You agree that (1) there were global catastrophes, and (2) that at least one of them occurred in the not too remote past. These conclusions will make you, too, a heretic in the eyes of geologists and evolutionists."

Were Velikovsky's theories really such nonsense? From the vantage of some fifty years later, when responsible scientists are talking about a Planet V that plunges into the Sun or about Uranus and Neptune migrating past Saturn, one has to wonder.

The main difference between Velikovsky's ideas and the computer models of Chambers, Levison, and others described in Chapter 9 is that the latter place their wandering planets in the early solar system, while Velikovsky placed his in the solar system of today. The difference is an important one: The early solar system was a dynamically "hotter" place that had not settled down yet into a near steady state. For Velikovsky's theories to be taken seriously, he would have to explain how Venus quietly stayed in orbit around Jupiter for four and a half billion years before suddenly breaking loose and going on its mad romp around the solar system. Chambers' and Levison's computer models also pay scrupulous attention to the principles of physics. Velikovsky did attempt to bring some physics into his models (that was one reason for consulting Einstein), but he was doing so *after the fact*—trying to fit the laws to his conclusions rather than deriving his conclusions from the laws of physics.

Nevertheless, I think that the violent reaction of scientists against Velikovsky was not based on such niceties. They were reacting against the *very idea* of Venus breaking loose from Jupiter, swooping past Earth, etc., not to mention the *very idea* of an outsider telling them how the solar system worked. Velikovsky's book embraced a style of discourse that scientists had long since discarded: a rhetorical style buttressed by citations of ancient texts. Four hundred years ago, Galileo had proved that quantitative reasoning based on experiments, not consultation of ancient authorities, was the gold standard for predicting how objects behave in the real world. Yet Velikovsky explicitly demoted the scientific method to second place: "If, occasionally, historical evidence does not square with formulated laws, it should be remembered that a law is but a deduction from experience and experiment, and therefore laws must conform with historical facts, not facts with laws." Velikovsky was asking physicists to abandon a method that defines their subject. It is no wonder they could not accept his proposal.

What is the relevance of the Velikovsky affair to the Moon's origin? Neither Velikovsky nor his followers ever proposed that the Moon originated in a collision or a close encounter between our planet and another. However, the episode abundantly illustrates the deep distrust that many scientists had for catastrophic scenarios in planetary science. Not only that, it probably increased their distrust. This was the hurdle of skepticism that the Big Splat had to clear. Before advocates of a giant impact could convince anybody that the theory was correct, they would first have to convince them that it was *plausible*. Not until 1974, after the Apollo missions had cast doubt on all the "classical" theories of the Moon's birth, did another scientist emerge who was willing to make the attempt.

Just Plain Smart?

Astronomer William Hartmann arrived as a graduate student in Tucson, Arizona, in the fall of 1961, perhaps the most propitious possible time for someone who wanted to study the planets. It was the dawn of the space age, and there were money and jobs galore for planetary scientists. For centuries, astronomers had studied the Moon simply because it was there. But now there was a new sense of purpose to their work: They were doing reconnaissance work for real explorers, who would walk on that Moon before the decade was out.

Hartmann came to study at Gerard Kuiper's brand-new Lunar and Planetary Laboratory, which had opened the previous year. Kuiper was deep into the project of mapping the Moon for the future astronauts. He assigned Hartmann the task of projecting photographs of the Moon onto a large, three-foot-wide globe, so they could see the Moon from various angles, as the astronauts would. Craters that are near the limb look very narrow from Earth, because we are seeing them almost edge on; but when Hartmann and Kuiper projected them on the globe, they became round and symmetrical. Most spectacular of all was Mare Orientale, a sea that is just barely visible to Earthbound observers. It is on the western edge of the Moon, just south of the equator. In fact, its center is normally beyond the Moon's horizon, but the libration (side-to-side oscillation) of the Moon sometimes brings it into view. Astronomers had long

Mare Orientale, as photographed by *Lunar Orbiter 4* in 1967. This spectacular "bull's-eye" basin proved that impacts could create even the Moon's largest features. Although Mare Orientale is just visible from Earth, on the Moon's western limb, its beautiful circular shape was not suspected until 1962, when William Hartmann and Gerard Kuiper projected photographs of it onto a globe. *Photograph courtesy of National Space Science Data Center and Leon Kosofsky.*

considered Mare Orientale to be just another irregularly shaped dark patch.

But from the side, it looked like a giant bull's-eye: a perfectly circular plain surrounded by three concentric rings of mountains. The beautifully preserved rings left little doubt of its origin: It was an impact crater the size of Texas. Hartmann and Kuiper estimated that the object that excavated the Orientale Basin had to be at least ninety miles in diameter. This wasn't just a rock that had slammed into the Moon, but a healthy-sized asteroid. (The largest object in the asteroid belt, Ceres, is about six hundred miles in diameter.) If the Orientale Basin had been excavated by such an encounter, it was likely that the other, less well-preserved basins had been, too—including the biggest one on the near side of the Moon, the Imbrian Basin.

It might seem like a short step from asteroid-sized impactors to planet-sized ones, but the time was not yet ripe. "In 1961, the whole emphasis was on looking at finer and finer detail on the Moon," Hartmann says. With bigger telescopes, astronomers could see smaller craters—say, half a mile to a mile across. Then the unmanned Ranger spacecraft crash-landed on the Moon, taking pictures all the way, and scientists could now see craters that were only a hundred yards wide. Giant craters were forgotten in the excitement over little-bitty craters. Even Hartmann worked for the next few years on crater-counting, a way of estimating the age of lunar features by counting the density of craters of various sizes. This was how he arrived at his correct estimate that the Moon's maria were about 3.6 billion years old.

But Hartmann did not stop thinking big, and this led him to another prediction that later seemed almost clairvoyant. In the words of geologist Don Wilhelms, "I have always thought that Bill has led a charmed life, although he may be just plain smart." Hartmann noticed that the highest mountains on the Moon's near side are on the very edge, near the South Pole. Mountains on the Moon usually occur around the edges of impact basins. (The rings of mountains surrounding Mare Orientale are the best example.) Therefore, Hartmann argued in 1962 that there must be an impact basin just over the horizon, on the back side of the Moon, larger and deeper than any of the impact basins on the near side. Hartmann called it the "Big Backside Basin." Between 1968 and 1970 the Soviets flew four unmanned spacecraft, *Zonds 5* through *8*, which took stereoscopic pictures of the far side and showed that Hartmann was right. The basin is now called South Pole-Aitken, referring to the features that mark its southernmost (South Pole) and northernmost (Aitken Crater) limits.

In the late 1960s Hartmann did one other thing that was either lucky or just plain smart: he started reading the papers of Safronov. At that time they were barely known in the West: "There was this feeling that Soviet astronomy was second-rate," Hartmann says.

Hartmann, however, was impressed by Safronov's power law indicating the size distribution of planetesimals, which meshed perfectly with his own ideas on the size distribution of craters. Safronov was the one other scientist who seemed to think that there was nothing unusual about hundred-mile-wide planetesimals careening into Earth or the Moon. But Safronov's models were entirely based on paper-and-pencil calculations. Hartmann decided to see what would happen if they were run through a computer instead. To do this, he went to a former classmate of his at Arizona, Don Davis.

By this time, the early 1970s, the intoxicating era of the space race was winding down. With money drying up, the scientists who had worked on Apollo were forced to go looking for other jobs. Davis had been a member of the Apollo team that worked on trajectories, de-

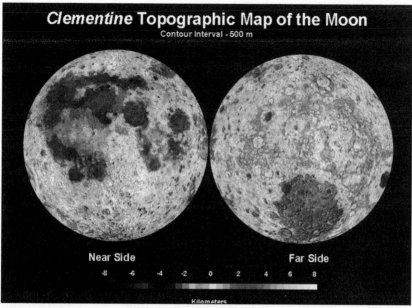

Clementine Topographic Map of the Moon
Contour Interval - 500 m

Near Side   Far Side

-8   -6   -4   -2   0   2   4   6   8

Kilometers

Lunar topography, as measured by the satellite *Clementine* in 1994. The near side is dominated by basins that have been filled in by lava to form maria. The far side has both the highest and lowest terrain on the Moon, and is dominated by the spectacular five-mile-deep South Pole-Aitken Basin. Curiously, this basin was never filled in by lava, so it does not stand out as conspicuously in photographs. The permanently shadowed craters where *Clementine* found water ice lie at its edge, near the South Pole. *Image courtesy of Lunar and Planetary Institute.*

signing the computer programs that told the astronauts when they needed to fire their engines. Their moment of glory had come on the ill-fated *Apollo 13* mission, when they had to plot a new course on the fly after an explosion had damaged the command module. Three astronauts' lives depended on the computer code working right, and there would be no second chance to check the answers. Says Davis, "It was a hell of a lot scarier than the movie."

After *Apollo 17,* Davis moved back to Tucson, where along with Hartmann and a few others he helped found a new think tank called the Planetary Science Institute. Using his expertise in celestial dynamics, he wrote a program to simulate the accretion of planetesimals in the early solar system. The results confirmed Safronov's theories, with one modification: The second-largest body in Earth's accretion zone had grown to a much larger size than Safronov had been willing to admit. It could even have been as large as the Moon itself.

Hartmann says he doesn't remember when this realization first dawned on him, but by 1973 he had a serious hunch: That second-largest object had not quietly gone into orbit around Earth. At some point it had plowed right into us, in a planet-shattering cataclysm that dwarfed the Orientale impact, the Imbrian impact, and even the South Pole-Aitken impact. Those collisions had merely thrown up mountain chains. Hartmann now envisioned an impact that had created the whole Moon.

Out of the Closet

Like Daly, Hartmann was very cautious about unveiling his idea in public. Even in the 1960s, he knew, his focus on large impacts had been out of step with the conventional wisdom. For a whole year he sat on his theory, asking geochemists whether the Moon rocks contained any clues that might point to a violent birth. Over and over they told him no. "I thought gee, these guys know what they're talking about. They measure rocks, and I just count craters," Hartmann says. So he held back until a meeting on planetary satellites at Cornell University in 1974.

In a short talk there, Hartmann outlined the basic arguments for the giant impact hypothesis. It solved the problem of why Earth has a large iron core and the Moon has almost none: in a giant impact, the iron in the impactor would have penetrated Earth's surface and stayed there. It explained the lack of volatile elements on the Moon: they would have boiled off in the explosion after the Big Splat. It explained why the Moon had a magma ocean, and it explained why the kinds of rock found on the Moon were similar to those in Earth's mantle: because a lot of the material in the Moon actually came from Earth.

In short, everything that seemed puzzling about the Moon in the context of the fission, capture, and coaccretion theories had a natural explanation as a consequence of the giant impact. But the same idea had foundered before, because it didn't pass scientists' sniff test for plausibility. Would it do so again?

At the end of the talk, one hand went up in the audience. It belonged to Alastair G. W. Cameron—"this big astrophysicist," Hartmann recalls. "When I saw his hand go up, I thought he was going to trash everything I said."

Instead, Cameron came out of the closet. He announced that he, too, had been studying the idea of a giant impact. For the last year,

he and a postdoctoral student named William Ward had been working on their own simulations of the event, and were convinced that it
was theoretically possible. "It was as if he had sprinkled holy water
on the idea," Hartmann says. Cameron had only one caveat: to eject
enough material into orbit to form the Moon, the impactor had to
be even bigger than Hartmann and Davis were proposing. It had to
be not just as big as the Moon (one-eightieth the mass of Earth), but
at least the size of Mars (one-tenth that of Earth).

Cameron had come to the question of the Moon's origins from a
completely different direction from Hartmann. Where Hartmann
had worked his way up from craters and meteorites, Cameron had
worked his way down from the stars.

Cameron made his reputation in the 1950s by explaining how elements heavier than helium are created in stars. Although in ordinary
English one often says that the Sun "burns," it doesn't really. The
Sun and other stars generate energy by nuclear fusion, the same
process that powers a hydrogen bomb. The main nuclear reactions
inside the Sun involve the fusion of atoms of hydrogen (the lightest
element, which has one proton) to form helium (the next-lightest
element, with two protons and two neutrons) through a series of
intermediate steps. Each time, the mass of the product is ever so
slightly less than the mass of the constituents. The mass difference is
converted to energy, according to Einstein's famous equation $E = mc^2$.
Because c, the speed of light, is a very large number, this equation
says that a little mass, m, is converted into a stupendous amount of
energy, E. And if you have a *large* amount of mass, as in the Sun, you
get enough energy to keep our planet warm for billions of years.

Later in a star's life, its temperature rises to the point where the
helium atoms start fusing to form carbon. This is the "red giant"
stage, after which the star typically collapses to form a white dwarf.
(This is the path our Sun is expected to follow.) But for a few stars—
those that are at least eight times larger than our Sun—a more spectacular fate awaits. The temperature and pressure just keep on rising,
and the carbon atoms fuse into even heavier elements: first oxygen,
then neon, and so on, all the way up to iron. Once the star gets
to that point, the jig is up. There is no way to generate more energy
by fusing iron atoms. (In fact, the reverse is true: fusing iron atoms
consumes energy.) With no more fuel to burn, the swollen star cannot maintain its size any longer. The core collapses, and a shock
wave blasts outward in a grand and glorious convulsion called a

supernova. Many cosmologists believe that the shock wave from such an explosion initiated the Sun's condensation from a gas nebula into a star. And the accumulated residue from many aeons of supernova explosions seeded the solar nebula with the heavy elements needed to make the planets.

By 1959, Cameron, using only a primitive calculator that belonged to the Accounting Department at the Canadian Atomic Energy Project in Ontario, had worked out the entire sequence of nuclear reactions up through the production of iron, including the temperature and pressure they occur at and the amount of energy they release. (Elsewhere, a team of four British and American astrophysicists— Geoffrey and Margaret Burbidge, William Fowler, and Fred Hoyle— was doing the same thing. Since they published first, their names are more commonly associated with this chain of nuclear reactions, now known as the "BBFH process.") At about this same time, Cameron began to turn his attention to the evolution of the solar system. If the solar system was made up of the residue from supernova explosions, it should be possible to estimate how common each element was in the beginning of the solar nebula. This would provide a starting point for the chemical history of the solar system.

Cameron's career, like Hartmann's, now took a detour that was strongly influenced by the space race. While the Americans were planning a manned mission to the Moon and unmanned missions to the planets, the Canadians were talking, at most, about sending a few rockets up to study the upper atmosphere. The right place for someone interested in the solar system was obvious, so in 1961 Cameron moved to the United States and began working for NASA. Later he took teaching positions at Yeshiva University and Harvard University, but he continued to advise the space program as chair of the National Academy of Sciences' Space Science Board until 1982.

Shortly after his arrival at Harvard in 1973, Cameron began to ponder the puzzle of the Moon's angular momentum. Earth and the Moon are like two skaters on an invisible sheet of ice, joined by an invisible thread, constantly rotating about a single stationary point called their center of mass. Like skaters, if they move toward each other they start spinning faster, and if they move apart they spin slower, but they are powerless to stop spinning entirely unless there is external friction between them and their environment. Real skaters do, of course, have friction to slow them down if they want. But

Earth and the Moon have nothing to rub against, so they keep on turning forever. Physicists call this fact the conservation of angular momentum.

Though it has one-eightieth of Earth's mass, the Moon contains five-sixths of the angular momentum of the Earth-Moon system, because the Moon is farther from the center of mass. To put it another way, if the Moon at any time in the past was part of Earth, then Earth must have been spinning six times faster—once every 4 hours—to maintain the conservation of angular momentum. It would have been by far the most rapidly spinning planet in the solar system. (Jupiter, the fastest, rotates once every 10 hours. Venus, the slowest, rotates once every 250 days.)

For the fission hypothesis, this was still not fast enough; the proto-Earth needed to spin even twice as fast as this to throw off part of itself into orbit. For the other theories, the angular momentum was too great. For example, it was hard to explain how randomly accreting planetesimals, some moving west to east and others moving east to west, could have produced the Moon's pronounced west-to-east motion.

Just as a thought experiment, Cameron asked himself whether the angular momentum of Earth and the Moon could have been provided by a single glancing blow from another object. Cameron calls it an "exercise in high-school physics" to figure out how large such an object would have to be, assuming it was moving at just Earth's escape velocity. The answer was about one-tenth the mass of Earth, or about the mass of Mars. If the impactor was traveling faster it would not need to be quite so big; on the other hand, if the impact were more head-on it would have to be bigger. Given the uncertainties, "the mass of Mars" was a good ballpark estimate, and easy to remember.

But working out the size of the impactor was still a long way from proving that such a collision would actually create the Moon. Cameron was back in the same position that Count Buffon had been in 1749, when he came up with the idea that the planets had been made by a collision between a comet and the Sun. Even if the collision between this Mars-size planet and Earth did throw chunks of Earth far out into space, they would not stay there unless something happened *after* the impact to round out their orbits. Otherwise, they would either escape Earth's gravity entirely or fall back to Earth.

Instead of an Earth and a Moon, we would simply have a larger Earth, a four-hour day, and no Moon.

But Cameron didn't give up. He and Ward theorized that the collision between Earth and the impactor had been so violent that it vaporized the impactor and part of Earth, too. A cloud of vaporized rock would behave differently in one very important way from an equivalent cloud made of solid chunks of rock. It would expand outward like the exhaust of a rocket. Because the pressure would make the gas expand uniformly in all directions, some of the cloud would get enough of a forward momentum boost to go into orbit. Once in orbit, the gases would cool and solidify into solid pieces of rock that would stay in orbit. The Moon would accumulate from this ring of solid debris.

Although the original angular momentum calculation was "high-school physics," this was way beyond that level. It all depended on the way that gases, liquids, and solids behaved at the very high temperatures and pressures created by the impact. Cameron and Ward had begun running simulations, but were not ready yet to announce their results, when Cameron went to the Cornell conference and heard Hartmann's talk.

The four modern originators of the giant impact theory—Hartmann, Davis, Cameron, and Ward—have never gotten into a dispute over who came up with the idea first, as often happens in science. One reason may be that each group benefited so much from the other group's argument. Hartmann and Davis had a reason for Earth to encounter a giant impactor, but no detailed explanation of how the collision could put material into orbit. Cameron and Ward, on the other hand, could explain what happened after the impact but did not specifically address why the impactor was there in the first place. Among the four of them, they had the makings of a plausible theory.

Interlude

Hartmann and Davis did, however, beat Cameron and Ward into print. Hartmann and Davis's version of the giant-impact hypothesis appeared in *Icarus,* a journal of planetary studies, in 1975, while Cameron and Ward published only a brief, three-page abstract of their work in 1976. *Icarus* was a far more appropriate venue for the new theory than the *Proceedings of the American Philosophical Society*, where Daly had published his little-known paper in 1946. *Icarus* is a journal that every planetary scientist reads, and it is unusual among scientific journals in specifically encouraging adventurous ideas. It is named, after all, not after Daedalus—the sensible father—but after Icarus, the reckless son who ventured too close to the Sun. Daly could not have published in *Icarus* because the journal did not exist in his time; it was founded in 1962, in the gung-ho era of the space race.

Even though it appeared in a well-read journal, Hartmann and Davis's paper was followed by a nine-year hiatus. Hartmann acknowledges that they could have done more to promote the idea, but he was already shifting his attention to Mars, which would be visited in 1976 by the Viking lander. A couple of other research papers during this interlude mentioned the giant impact theory, but most of the attention was going to increasingly elaborate versions of coaccretion. Ted Ringwood, an Australian geologist, gave the Big Splat a qualified endorsement in his 1979 book *Origin of the Earth and Moon:* "The author now believes that the answer to the Moon's origin is to be sought in the general direction pointed by Hartmann and Davis, and by Cameron and Ward." And: "They rely on processes that must *inevitably occur* during the accretion of planets and can supply the enormous amounts of energy required." However, he continued to hold out for several macroimpacts to build up a ring of debris about Earth, rather than one megaimpact.

Meanwhile, Cameron continued to work on his computer simulations, but they were still very crude. One thing that he discovered during this time was that the expanding gas idea didn't work. There wasn't nearly enough energy in the giant impact to vaporize the whole impactor. Most of the debris was solid, and another, quite unanticipated effect, called "gravitational torque," was keeping it in orbit after the giant impact.

Gravitational torque is a bit tricky to understand. But we probably would not have a Moon without it, so it's probably worth the trouble. Conventional wisdom, since the time of Laplace, holds that if you launch an object from Earth with one big thrust—like shooting it out of a cannon—then it has only two options: it can escape Earth's gravity, or it can fall back to Earth. It *cannot* go into orbit. However, there is a hidden assumption behind that conventional wisdom: it assumes a spherically symmetric Earth.

William Hartmann, coproposer of the giant impact theory of the Moon's creation, works at his art studio in Tucson, Arizona. Hartmann devotes as much time to his art as he does to his scientific research; his specialty, of course, is space art. He believes that his artist's eye and his penchant for looking at the big picture have helped him make discoveries that other scientists missed—including, of course, the giant impact theory. *Photograph copyright © by Dana Mackenzie.*

Immediately after the Big Splat, Earth was nowhere near spherically symmetric. It looks in Cameron's pictures like a cosmic amoeba. The hemisphere that the impactor didn't hit is still more or less intact, but the impact side has been vaporized, liquefied, and pulverized to the point where you cannot tell what shape it started out in. Meanwhile, the impactor has been sheared by the impact so that it forms a long, narrow arm extending well beyond the planet's surface.

Now Earth is rotating faster than the arm of material from the impactor. This means that the bulges of the now asymmetrical Earth get ahead of the arm, and they tug it forward gravitationally. (This should remind you of Darwin's tidal friction, tugging the Moon forward in its orbit.) Instead of just going up and then down, the debris are getting pulled *around* Earth. That is the gravitational torque. Later, Cameron's simulations would show that the arm of debris gets pulled into a spiral, and that the spiral arm itself greatly enhances the gravitational torque. It acts like a siphon, lifting the post-impact debris into a higher and higher orbit.

While Cameron worked on his computer, Hartmann was trying to visualize the Big Splat and its aftermath through art. The grandson of a painter, he had grown up in a house full of paintings. For many years he had let his talent for painting lie dormant, but he returned to it when he needed illustrations for his first astronomy book. He still spends as much time in his art studio as in his office, and believes that his artist's eye is an important contributor to his science. It forces him to look at the big picture in a way that science does not. "Scientists are taught to be analytic, to analyze one phenomenon at a time," he says. "An artist has to synthesize, to ask: What color is the sky? The light? The rock?"

Hartmann's paintings of the giant impact have appeared in *Astronomy*, *Natural History*, and the *New York Times*. They are still the only widely available portraits of the doomed planet that was annihilated in the collision, and of the newborn Moon that formed from its ashes. Based on the best computer models available at the time, they speak in a language that everyone can understand—not a language of particles and hydrodynamics, but a language of fire and light and darkness. Perhaps as much as Cameron's calculations, Hartmann's paintings helped keep the Big Splat alive during a time when most scientists were still not paying attention.

11

The Kona Consensus

It would be hard to find a more appropriate place to hold a conference about the Moon than Kona, Hawaii. As your plane from Hono-
lulu begins its descent, you're thinking of swaying palms, tropical
breezes, sandy beaches . . . but suddenly you look down, and all you
see is lava—black, solidified lava, with not a single living thing growing in it. Your first feeling is one of disorientation: Wait a minute.
Where am I? Am I still on Earth or did I get snatched away to another
planet somehow?

Of course, after you land and get to the touristy parts of town,
you do see palm trees and beaches. But the scenery on the approach
to the airport is a reminder that there is another side to the big
island of Hawaii. Not far away, in Hawaiian Volcanoes National Park,
the landscape looks just like a moonscape, stark and simple. This is
a place where astronauts even came to train for the Moon missions,
so they could learn how to identify volcanic rocks.

Early in 1983, Bill Hartmann, Roger Phillips (director of the Lunar
and Planetary Institute in Houston), and Jeff Taylor (a geologist at
the University of Hawaii) decided to organize a conference devoted to
the origin of the Moon. It would be held in the resort town of Kona.
More than a decade had passed since the last Moon landing, and it
was high time to face the biggest mystery left from the Apollo years.
They sent out a challenge to their fellow researchers: You have eighteen
months. Go back to your Apollo data, go back to your computer, do
whatever you have to, but make up your mind. Don't come to our
conference unless you have something to say about the Moon's birth.

Hartmann says he never thought that the giant impact model
would come out on top in the debate, only that it would get some
attention. But in the summer of 1984, he and Taylor got together to
read the abstracts that other scientists were sending in, one- to two-

page digests of the talks they were planning to give. It was like getting the first exit polls from an election. Hartmann and Taylor could tell then that a big upset was brewing, but no one else knew it until October.

Once in a Lifetime

To understand what was so special about the 1984 Kona meeting, why it is almost legendary in the lunar science community, you first have to realize one thing about scientific meetings: nothing ever gets decided at them.

At a typical scientific meeting, everybody knows beforehand who the main players are and what their opinions are. The talks are prepared long in advance, and there is nothing spontaneous about them. Every now and then a speaker might spring a surprise—a new finding or a new theory that genuinely changes things—but even then, it doesn't usually change the field right away. The other specialists have to go home and process the new information. Old theories have to be sifted through and reappraised. More papers come out in favor of the new hypothesis, and others come out against it. Eventually, sometimes after many years, a new consensus emerges.

Kona was a once-in-a-lifetime conference where the attendees actually changed their minds. Not all of them, of course, but enough of them to create a new consensus almost overnight. Don Wilhelms, who attended the meeting, describes it well in his book *To a Rocky Moon:*

> The conference was incredible. Outside the hotel's conference room were the beaches and soft climate that most people find appealing. But nobody stirred. In anticipation of the usual inconclusive hypothesizing, the organizers had entitled two conference sections "My Model of Lunar Origin I" and "My Model of Lunar Origin II." But to everyone's surprise, "our model" emerged from the presentations of one speaker after another.
>
> With great relief most of the conferees discarded the traditional theories in their original forms.

Ever since Kona, the field of selenogony—the origin of the Moon— has been radically realigned. Before the conference, there were partisans of the three "traditional" theories, plus a few people who were starting to take the giant impact seriously, and there was a huge apathetic middle who didn't think the debate would ever be resolved. Afterward there were essentially only two groups: the giant impact camp and the agnostics.

For the most part, the change in attitude came about not because of a single revelation, but because so many of the speakers, when forced to commit themselves, had concluded that the giant impact looked better than any of the alternatives. And they had done so independently. Until the meeting, only Hartmann and Taylor had realized how strong a consensus was emerging. For everyone else it was a surprise, and Hartmann recalls seeing people walking around with dazed looks on their faces.

If there was any talk that made a big impression at Kona, it was probably Hartmann's own lecture, called "Stochastic Does Not Equal Ad Hoc." In this talk, he addressed the main scientific complaint against the giant impact theory: it seemed too "ad hoc," or what I have been calling in previous chapters a "just-so story." It violated the scientist's taboo against unique, one-of-a-kind events. Scientists hate flukes.

The point of Hartmann's talk, and a vision that he still enunciates with great fervor, is that the Big Splat was *not* a one-of-a-kind event but simply the largest in a spectrum. In the early solar system, there was a continuous size distribution of objects—ten Moons for every Mars, ten Marses for every Earth, and so on. (The specific proportions are still somewhat uncertain, but the general form of the relationship between size and abundance, a "power law," is not.) That same distribution will reflect itself in the impact history of every planet. For every hundred crater-scale impacts, there will be a basin-scale impact. For every hundred basin-scale impacts, there will, in all likelihood, be one planet-scale impact.

But when you are dealing with the single largest event, the stochastic (random) nature of the impact process really manifests itself. Will the impact be on center? If so, there will be no Moon, and you will be left with a very slowly rotating planet, such as Venus. Will it be glancing? Then you might get a large Moon and a hefty dose of angular momentum, as Earth apparently did. Will the impactor hit the planet near one of the poles? If so, you might get a planet tipped on its side, such as Uranus. Yes, the Big Splat was a random event, and its details could not have been predicted in advance. But no, it wasn't a fluke. Events such as the Big Splat are as foreseeable as automobile accidents on a highway full of speeding commuters.

   In addition to this argument, I believe that psychological factors may have helped the Big Splat finally get over the plausibility hurdle. In the early 1980s, the barriers to catastrophic theories were coming down all over science. In 1980, Luis and Walter Alvarez announced their theory that a huge meteorite had struck Earth sixty-five million years ago, precipitating an ecological disaster that killed the dinosaurs. (They had discovered a worldwide layer of iridium at the top of the same geological stratum that marks the final appearance of the dinosaurs. Because iridium is a siderophile [iron-loving] element, and Earth's crust is siderophile-poor compared to primitive meteorites, the iridium-rich layer most likely contains extraterrestrial material.) This fits in perfectly with Hartmann's theory: the meteorite that killed the dinosaurs was another of those random but inevitable events. Even though its effects were felt around the globe, it was about ten million times smaller than the giant impactor that created the Moon—seven rungs below it on the scale of catastrophes. Back in the early days of Earth, or in the days of the late heavy bombardment if you subscribe to that theory, such an impact may have occurred roughly every twenty years!

   A similar "paradigm shift" was brewing in the rarefied world of pure mathematics, a world that practical planetary scientists rarely paid any attention to. This change had its roots in celestial dynamics, with the efforts of mathematicians such as Henri Poincaré to make sense of the problem Isaac Newton hadn't been able to solve: the motion of three objects that act on each other solely by the inverse-square law of gravity. Poincaré had discovered, back in the 1880s, that in some cases the motion of the three bodies seemed to be inherently unpredictable. Even the tiniest change in their initial positions can make their long-term history totally different. But this concept was utterly alien to the nineteenth-century scientist and his deterministic, clockwork view of the universe. Even Poincaré was unable to put his discovery into words. It was only in 1974 that a mathematician named Robert May came up with a name for it: chaos.

   The story of the emergence of chaos into a coherent field of science has been told in James Gleick's 1987 best-seller *Chaos,* and I will not retell it here. It will suffice to say that over the course of a decade scientists in many disciplines, from weather forecasting to population biology, realized that deterministic systems could behave in a seemingly random way. As early as 1981, an astronomer named Jack Wisdom brought the idea of chaos back into planetary science where it started, by showing that the tumbling motions of an irregularly shaped asteroid were chaotic.

I am told by Bill Hartmann that chaos theory had nothing to do with his proposal of the giant impact, or with the discussions at Kona. If so, it was an opportunity missed. Chaos theory could have provided a philosophical foundation for believing what expensive and time-consuming computer calculations would ultimately prove: that the early solar system was a very dynamic and unstable place, a giant shooting gallery in which big collisions could occur, planets could migrate, and so on. If the conferees at Kona were not influenced consciously by chaos theory, then perhaps it was some sort of collective unconscious that made them ready, like scientists in so many other fields, to let a little unpredictability creep into their view of the world.

Making the Grade

After Kona, the giant impact theory may have been winning the battle for plausibility, but that did not necessarily mean it was correct. It had to go through the same scrutiny and comparison with known information about the Moon that the other theories did. The Kona conference started that process but did not end it.

On the last day of the Kona meeting, John Wood prepared a "report card" for each of the major theories. He assessed how well each theory performed on eight "tests":

1. *Lunar mass.* Can the theory explain why we have such a large Moon?
2. *Earth-Moon angular momentum.* Does the theory agree with the rate of rotation of Earth, and rate of revolution of the Moon?
3. *Volatile element depletion.* Does the theory explain why elements with low boiling points are much less common on the Moon than on Earth?
4. *Iron depletion.* Does the theory explain why the Moon has such a small iron core compared to Earth?
5. *Oxygen isotopes.* Does the theory explain why the Moon's oxygen isotopes match Earth's?
6. *Similarity of trace element patterns.* Does the theory explain Earth and the Moon's similar pattern of siderophile depletion? (Wood put these grades in parentheses because he was not convinced that the observed similarities were really reliable.)

7. *Magma ocean.* Is the theory consistent with a magma ocean early in the Moon's history?

8. *Physical plausibility.* Is it a "just-so story"?

Wood graded five different theories, including "disintegrative capture," a new version of the capture hypothesis in which the captured planet breaks apart, and part of it goes into orbit around Earth while the rest escapes Earth's gravity. Here is his report card:

	Capture	Coaccretion	Fission	Giant Impact	Disintegrative Capture
Lunar mass	B	B	D	I*	B
Angular momentum	C	F	F	B	C
Volatile depletion	C	C	B	B	C
Iron depletion	F	D	A	I*	B
Oxygen isotopes	B	A	A	B	B
Trace elements	(C)	(D)	(A)	(C)	(C)
Magma ocean	D	C	A	A	B
Plausibility	D−	C	F	I*	F

* I — Incomplete

Although this report card reflects only one person's opinion, it provides a pretty clear picture of why the giant impact theory came out on top. Each of the other theories failed at least one important test, and there was a liberal sprinkling of D's as well. Only the giant impact theory was consistent with everything we knew about the Moon. And lunar scientists were especially happy with it because it explained the major discovery of the Apollo missions, the lunar magma ocean.

On the other hand, Wood's report card was not exactly a ringing endorsement. The I's showed a number of "assignments" that the Big Splat's proponents still had to complete. They would have to strengthen the plausibility argument. The problem was not so much the plausibility of the impact itself, but whether it could lift enough debris into orbit to form the Moon, and whether that material would in fact accrete into a satellite, or stay in a Saturn-like ring, or eventually fall back to Earth. Also, they would have to prove that the impactor's iron would go into Earth's core, not into orbit.

One other question doesn't show up on Wood's report card but has turned into a real sticking point for geochemists since 1984. That is the consistency of the giant impact with *Earth's* history. The giant impact would almost certainly create a magma ocean on Earth as well as on the newborn Moon, but there is little to no evidence that such a magma ocean ever existed. Of course, we have little knowledge of *anything* that happened on Earth in the first billion years of its existence, because the evidence has been destroyed by plate tectonics.

The debate over whether Earth had a magma ocean is still going on. One geochemist, Michael Drake of the University of Arizona, started out as a skeptic but finally convinced himself a few years after Kona that there really was a magma ocean on Earth. Others have remained skeptical, including John Jones of NASA's Johnson Space Center, who theatrically exclaimed at a follow-up meeting in Monterey in 1998, "God, I hate giant impacts!"

The Digital Cosmos

If the giant impact hypothesis had come along thirty years earlier—as it did, if one counts Daly's single paper of 1946—it would have been impossible to evaluate properly. Only the invention of the computer has made it feasible to study the impact process with any degree of certainty, because an encounter of this scale can never be simulated in a laboratory.

Bill Hartmann was never a computer guru, and has essentially bowed out of the further development of the giant impact theory. However, Al Cameron continues to work on it. His small office in Arizona (where he moved after he retired from Harvard) is literally crammed with computers. He has four servers running around the clock; it takes about a month of server time to model a week of "real" time during the giant impact. After finishing a run, Cameron will feed the computer slightly different conditions to start with—different masses for Earth and the impactor, or a different angle of impact—and set it to work again. It's an incredibly time-consuming process. Sitting in Cameron's office, lit by the artificial glow of computer displays, one cannot help being reminded of George Darwin, sitting by the fireside in his comfortable house in Cambridge and toiling away for twenty years on his pear-shaped figures of equilibrium. "It's dogged as does it."

Lately Cameron has had some help and some competition in his lonely simulation game from Robin Canup, a young planetary scientist at the Southwest Research Institute in Boulder, Colorado. Her participation has been welcomed by others, too, because it is always dangerous to rely on one scientist's work alone. Good science should always be "replicated," or duplicated by an independent investigator, and the subject always benefits from the inevitable differences and exchanges of opinion that result.

Both Canup and Cameron use the "smooth particle hydrodynamics" (SPH) method, which also has been used to model bomb explosions and the disintegration of Comet Shoemaker-Levy. In this method, both Earth and the impactor are divided into "particles"—the more the better, but not too many for the computer to handle. The word "particle" is truly a misnomer, because each one is a sphere hundreds of miles in diameter. The particles do not have to be solid. A table of experimental data called the "equation of state" helps the computer determine whether each particle is in a solid, liquid, or gaseous state and what its pressure is for any given temperature and density. This is important, because explosive expansion of vaporized rock could boost some material into orbit (Cameron's original idea).

SPH programs run in a series of time steps corresponding roughly to two minutes of real time during the collision. At the beginning of each time step, the computer "knows" the density, energy, and momentum of each particle. Then it works out what each particle will do over the next two minutes—how much momentum it will lose or gain through collisions (if the particle is liquid or gaseous, the "collisions" are really changes in pressure) and how much energy it will lose or gain by gravitational interactions. Since every one of several thousand particles interacts gravitationally with every other one, this is an extremely involved calculation. After working out each particle's change in momentum and energy, the program updates their positions and densities, then starts over again with the next time step.

Cameron began using SPH shortly after the Kona conference, and it represented a major increase in sophistication of the computer models. It was now possible to track the movement of iron and rock separately, by identifying each particle as one or the other and using different equations of state for the two kinds of particles. In theory, "particles" of water or gas could be added to simulate the effect of the giant impact on Earth's water and atmosphere. However, neither Cameron nor Canup have attempted this refinement yet.

Broadly speaking, the simulations have filled in the "Incompletes" on the theory's transcript. First, they make it clear that under the right circumstances the impactor's iron does get swallowed up by

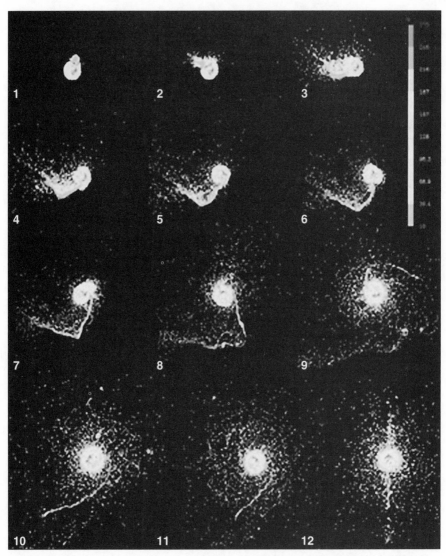

The giant impact, as modeled on a computer by Robin Canup and Erik Asphaug in 2001. After the initial impact and explosion (frames 1 to 3), Earth is significantly flattened (frames 4 and 5), and the "dent" is rotating ahead of the debris. The gravitational torque from this lopsided Earth boosts the debris into a higher and higher orbit and stretches it out into a long spiral arm (frames 6 to 8). Part of the arm crashes back into Earth, creating a second dramatic explosion that throws a more uniform—and hotter—disk of debris into orbit (frames 8 and 9). Secondary spiral arms form in this ring (frames 9 to 11), and these arms also help to boost the debris to a higher orbit. Frame 12 shows an edge-on view of the disk twenty-four hours after impact. *Image courtesy of Southwest Research Institute.*

The accretion of the Moon, as simulated by Eiichiro Kokubo of the National Astronomical Observatory of Japan and Hitoshi Miura of Musashino Art University. (a) This simulation begins roughly where Canup's leaves off, with two Moons' worth of debris in orbit around Earth. (b) The disk begins clumping up into spiral arms. (According to Canup's simulation, this probably would have happened already.) (c) One of the clumps (upper left) grows larger than the others, and runaway accretion begins, thanks to gravitational focusing. (d) Thirty days after impact, the biggest clump is beginning to look like the Moon. Twenty days later, it moves out past the Roche limit (the second circle drawn around Earth), and its survival is assured. *Images courtesy of Eiichiro Kokubo and Hitoshi Miura.*

Earth. Models of the postimpact accretion disk, by Canup and two Japanese scientists, Eiichiro Kokubo and Shigeru Ida, have showed that a uniform disk of material could condense into the Moon in just a year's time, with about 50 percent of the material in the disk falling back to Earth. So if the giant impact threw at least two lunar masses of debris into orbit, it would pass Wood's "lunar mass" test.

However, it would be premature to replace Wood's Incomplete grades with straight A's. (Wood himself tells me that he would now replace them with B's.) Cameron found that in a "late" Big Splat, with Earth almost completely formed, he could not get two lunar masses into orbit without exceeding the angular momentum limits or making the Moon too rich in iron. This led him to suggest that the Big Splat might have occurred earlier, when Earth was only half formed. He also upped the size of the impactor to two Mars masses. With a smaller Earth to escape from, it would not take quite so much angular momentum to pry the debris from the impact into orbit.

When Cameron tried to peddle this thesis at the Monterey meeting in 1998, he encountered a lot of resistance, because the Moon also would accrete a significant part of its mass *after* the Big Splat. The material accreted during this final stage would have Earthlike amounts of iron and volatile elements. This would weaken two of the selling points for the giant impact theory, namely its ability to account for the lower levels of iron and volatiles in the Moon's crust. (In rebuttal, Cameron argues that relatively little material would have stuck to the Moon in the first few million years after its birth, because its orbital velocity exceeded its escape velocity. This means that most meteorites that struck the Moon would be going so fast that their fragments after the collision would be launched back out into space, instead of staying on the Moon.)

Thus it was a relief to some people in 2001 when Canup, together with Erik Asphaug of the University of California at Santa Cruz, rehabilitated the original version of the theory, a late impact by a roughly Mars-size planet. In an article published in 2001, they said that the problems Cameron encountered may have been caused by using too few particles in his earlier simulations. For example, if Earth is represented in the computer simulation by four thousand particles, then the post-impact Moon (at one-eightieth of Earth's mass) would be represented by only fifty particles. Then the Moon's entire iron budget—estimated at 2 to 4 percent of its mass—would amount to only one or two particles! Thus if even one particle ended up out of place—due to round-off error in the computations, for example—it could make a whole run look invalid. Clearly, forty thousand particles (the size of the models Canup and Cameron are running now) would be an improvement over four thousand, and it might even take four hundred thousand (which is beyond the current limits of their computers) to be really confident of the results.

The Chemical Conundrum

Meanwhile, one of the long-running battles of the post-Kona era has been fought by geochemists over the implications of the giant impact for Earth. As I have mentioned, the abundance of siderophile elements in Earth's mantle is lower than in meteorites, which is a bit of a paradox if Earth accreted out of primordial material similar to that of the meteorites.

To be more precise, the siderophile elements have a distinctive distribution that many geochemists have interpreted as a clue to the way Earth formed. A graph of the distribution resembles a staircase with three steps, corresponding to three groups of elements. Slightly siderophile elements (such as manganese) are only slightly less abundant in Earth than in meteorites, moderately siderophile elements (such as silver) are more severely depleted, and highly siderophile elements (such as gold) are much rarer in Earth than in meteorites. The puzzle, though, is that the moderately and highly siderophile elements are not even close to being as rare as they should be (a lucky thing for silver and gold prospectors).

Chemists determine how rare the elements "should be" by laboratory experiments based on the assumption that these elements alloyed with iron and followed it into Earth's core. Gold, for example, will alloy with iron at a concentration seventy million times higher than it will dissolve into liquid rock. So the vast majority of Earth's gold should be in the core, and only a trace amount should be left in Earth's mantle. But in fact, gold is thousands of times more common in Earth's mantle than one would expect.

In 1981, Heinrich Wänke of Germany proposed an explanation that satisfied most geochemists but does not sit well at all with the idea of a magma ocean on Earth. In Wänke's model, Earth accreted in three stages. First came a metal-rich stage, when most of the core was formed. In the second stage, when the mantle was formed, the material forming Earth included more oxygen. This changes the "iron-seeking" behavior of the siderophiles, because more of the iron in this stage would combine with oxygen instead of remaining in its pure metallic form. Only the highly siderophile elements would continue to alloy with iron and precipitate into the core; the moderately siderophile elements would not, and they would remain trapped in the mantle. Finally, in the third stage, the accreting material becomes so oxidized that it doesn't contain any metallic iron at all. In this "late veneer," representing maybe 1 percent of Earth's mass, even highly siderophile elements such as gold will stay near Earth's surface, and that is why we have gold deposits today.

This three-layer cake baked by Wänke would, of course, be totally ruined by the Big Splat. The carefully separated layers would be stirred up into one uniform batter, the magma ocean. This did not mean that Wänke's idea of "heterogeneous accretion" was wrong; according to Michael Drake, it's almost a certainty that the proto-Earth's chemical environment did change as it was accreting. But it does mean that heterogeneous accretion cannot be used as an explanation of the discrepancies in the concentrations of iron-loving elements. By 1988, when a Kona-like conference was held in Berkeley to discuss the origin of Earth, chemists had no idea what to believe. "I think I heard more ideas about the behavior of the early Earth's interior than there were geochemists there," says Al Cameron. Needless to say, no Kona-like consensus came out of this meeting.

But in the past decade the puzzle of the "excess siderophiles," as Ted Ringwood called it, started to be resolved. Before 1990, chemists had always determined the iron-seeking property of elements at "low" pressures (i.e., ordinary atmospheric pressure) and "low" temperatures (about 1,250 degrees Kelvin or 1,800 degrees Fahrenheit). These conditions were irrelevant for the bottom of a magma ocean, which would be hotter and at a higher pressure, but they were the only conditions that conventional furnaces could deal with.

However, in the early 1990s the prices of diamond anvils and multi-anvils used to re-create the high pressures inside Earth came down to the point where university laboratories could afford them. For the first time, chemists could experiment with siderophile elements at the actual temperatures and pressures of a molten ocean—and they found that many of the discrepancies simply vanished. For example, gallium switches from iron-loving to rock-loving at high pressure, which explained why that element—unlike the other siderophiles—is not depleted at all in Earth's mantle. Other siderophile elements become much less attracted to iron. Best of all, the discrepancies went away in a consistent manner. Drake, along with his University of Arizona colleague, Kevin Righter, found that the observed siderophile concentrations in Earth rocks would match experimental predictions if the magma ocean were at a temperature of 2,225 degrees

Kelvin (3,500 degrees Fahrenheit) and if the pressure were about 270,000 atmospheres. This would correspond to a magma ocean at least four hundred miles deep!

Many details remain unclear in the chemical history of early Earth. For starters, it is not exactly clear what Drake and Righter's estimates mean. Was there only one magma ocean that reached this temperature and depth at one time? Or does this represent an average over a long period? Could there have been several melting episodes due to several major impacts?

Also, much is still unknown about the behavior of siderophile elements at high temperature and pressure. There are perhaps ten laboratories performing these experiments, working their way through the periodic table one element at a time. In 1993 only the elements nickel and cobalt had been studied under magma-ocean conditions. By 1998 the list was up to seven elements. Perhaps there will still be some surprises when the chemists get to the highly siderophile elements. But for the time being, the idea of a magma ocean seems to be—to use a very inappropriate metaphor—on solid ground.

A Matter of Timing

Computer models can tell us *why* the giant impact occurred. (Any accreting planet is likely to be hit at least once by an object one-tenth its size.) They can tell us *how* the impact formed our Moon. (By throwing off a ring of debris, which was boosted into orbit by the "gravitational torque" effect and perhaps a little bit by the expansion of gas.) Geology and chemistry can tell us *what* happened next. (Magma oceans on both Earth and the Moon.) But there is one important question left: *When* did the Big Splat happen?

There are some basic constraints. The age of the oldest known materials in the solar system ("calcium-aluminum-rich inclusions" in meteorites) is about 4.57 billion years. This is often taken as Time Zero for the formation of the solar system. The oldest dated Moon rock, an anorthosite collected by *Apollo 16*, is about 4.44 billion years old. So the Moon certainly formed within 130 million years of Time Zero. But Safronov's Earth accretion models show that Earth should have finished accreting within 100 million years. This has been confirmed in computer models; in fact, Earth should have reached 90 percent of its mass within 40 million years from Time Zero. Assuming the giant impact was the last major event of Earth's accretion, it seems likely that the Moon formed 40 million to 100 million years after Time Zero. If, however, Cameron is right that Earth was only half complete at the time of the giant impact, then it would have taken place roughly 10 million to 20 million years after Time Zero.

Most of the ages of Moon rocks have been deduced through the analysis of radioactive elements with long half-lives, such as rubidium 87, which decays into strontium 87 with a half-life of 49 billion years. This means that a rock will lose half of its original composition of rubidium 87 in that time. Of course, Moon rocks are not anywhere near that old. If a Moon rock had, say, 95 percent of its original complement of rubidium 87, then its age would work out to be 3.6 billion years, typical of a mare basalt.

You might have noticed a flaw in this dating scheme. How can you tell what the "original" amount of rubidium 87 in a rock is? You can't go back in a time machine and measure the amount of rubidium when the rock first crystallized from the magma ocean.

The actual technique that isotope chemists use is a bit more complicated and clever. It starts by taking *several* different mineral samples from the same rock. (This is not hard, because all rocks, when looked at under a microscope, are a mixed-up jumble of different minerals.) All the samples presumably are the same age, but they have different initial rubidium concentrations, because rubidium has a different affinity for different minerals. Over the aeons, the crystals that had a lot of rubidium to start with will accumulate a larger fraction of strontium 87 than the ones that started with a small amount of rubidium. When the concentrations of rubidium and strontium 87 in the different samples are plotted on a graph, they form a straight line called an "isochron" (meaning "equal age"). The older the sample is, the steeper the line is, and vice versa. So the chemist can use the slope of the isochron to determine the age of the rock—in other words, the amount of time since it last melted. Melting resets the slope of the isochron to zero because it lets the excess strontium escape the mineral crystals and be replaced by rubidium again.

As a rule of thumb, a radioactive dating system is reliable for only about five half-lives of the "parent" element, because after that time not enough of the parent will remain to be measured accurately. That is why geologists usually have to use elements with very long half-lives, such as rubidium 87, to figure out the age of a rock. The most famous radioactive dating technique, carbon 14 dating, is virtually useless for geologists. Because carbon 14 has a half-life of 5,730 years, it cannot be used reliably to date objects more than 28,000 years old. That is fine for dating human artifacts, which are almost always younger than this, but not for dating rocks.

In spite of the five-half-life rule of thumb, the most exciting recent developments in lunar chronology do involve elements with shorter half-lives, such as hafnium 182, which decays to tungsten 182 with a half-life of 9 million years. Hafnium 182 is an "extinct radionuclide," an isotope that no longer exists naturally because it decays so rapidly. But it does leave a fingerprint. Any time an isotope chemist finds an atom of tungsten 182, he knows that it came from a hafnium 182 atom, because the "normal" flavors of tungsten have atomic weights of 183, 184, and 186.

Now, hafnium is a rock-loving element, while tungsten is metal-loving. Thus the decay of hafnium 182 can be used to tell when a planet's core formed. To see why, consider two possibilities.

1. The core formed early, when hafnium 182 was still "alive." In this case, rocks in the mantle would become enriched in hafnium 182, because hafnium is rock-loving; they would be depleted in tungsten 183, 184, and 186. As the hafnium decays, it would be replaced by tungsten 182. But this tungsten would stay in the mantle, because the iron it likes to alloy with has already fled to the planet's core. As a result, the planet's mantle would have a disproportionately high amount of tungsten 182 compared to the ordinary isotopes.

2. The core formed late, more than 45 million years (or five half-lives of hafnium 182) after Time Zero. In this case, all the hafnium 182 would be converted to tungsten 182 before the core formed. This tungsten would escape into the core along with the ordinary isotopes, because *chemically* they are no different from one another. (The missing neutrons in tungsten 182 make it lighter but do not affect its iron-loving properties.) Thus only a small amount of tungsten would remain in the mantle, and there would be no excess of tungsten 182.

To state things very succinctly, radioactive dating techniques that use long-lived elements, such as strontium, measure the time between a rock's crystallization and the *present.* Dating techniques that use short-lived elements, such as hafnium, measure the time between the rock's crystallization and the *beginning of the solar system.*

At the Monterey conference in 1998, a team of three isotope chemists led by Alex Halliday of Switzerland reported an amazing discovery: rocks from Earth's mantle do not show any tungsten 182 enrichment, but many of the Moon rocks do! In a meeting report, Frank Podosek of Washington University in St. Louis wrote, "If the moon were an independent planet, we would likely conclude that it was older than Earth." Taken at face value, Halliday's data suggested that Earth was born (or at any rate, its core formed) at least 50 million years after Time Zero, while the Moon's core formed 25 million years earlier. But how is this possible if the Moon formed from Earth in a giant impact?

At this point the answer is far from clear, but one possibility is particularly tantalizing. The hafnium-tungsten clock could be recording the time of core formation in the impactor *before* the Big Splat. This would be consistent with Cameron's computer models, which show that most of the rocks in the Moon came from the impactor, not from Earth. Thus it is quite possible that these rocks "remember" where they came from. If so, the hafnium-tungsten ratios may be providing us our first glimpse at the planet that is no more, the planet that made our Moon. You may be forgiven for thinking that it is a pathetic little scrap of information—just a date, nothing more. But like a tattered picture of a long-dead relative, it is all we have.

Synthesis or Revolution?

Before leaving the "Kona consensus," it is important to acknowledge that there are still skeptics among lunar scientists in America, though they are distinctly in the minority. The prediction of a magma ocean on Earth is still the weakest link in the theory, because all of the evidence for such an ocean is indirect. Some geochemists, such as John Jones and Timothy Grove, point out the existence of minerals from early in Earth's history that could only have formed if Earth's surface was cool enough to allow water oceans. In 2001, an Australian scientist named Simon Wilde set a new record by discovering a tiny zircon crystal, only as wide as two human hairs, that was 4.4 billion years old. But proimpact theorists such as Drake are not flustered. They argue that Earth's surface could have cooled in fewer than 100 million years. In fact, it could even have cooled in 1,000 years. Thus there could have been a magma ocean 4.5 billion years ago and a water ocean, in which the zircon formed, 4.4 billion years ago.

It's also obvious that computer models of the giant impact need to get better. Canup's and Cameron's simulations are still simplistic as far as the material makeup of Earth is concerned. They do not indicate how water or a primitive atmosphere might have reacted to the impact. There has been some debate about the validity of the "equation of state" used in the smooth-particle hydrodynamics code. Jay Melosh of the University of Arizona believes that the gases produced by the impact may be more explosive than we thought—which would be good news for impact theorists, because it would put more material into orbit.

Also, computer models of accretion in the solar system, and the accretion of Earth's postimpact disk into the Moon, are still far from perfect. No one really understands how planetesimals stick together. It is quite possible that the present models overstate their tendency to adhere and understate their tendency to break apart in collisions. This might make both planetary accretion and the Moon's accretion after the giant impact take longer. Finally, in solar system models, the giant impact generally leaves Earth in too eccentric an orbit—that is, its orbit around the Sun is too elongated compared to its current, nearly circular path. It remains unclear whether this is a defect in the computer models ("a very stubborn defect," according to John Chambers) or in the theory itself, as Evgeniya Ruskol argues.

The Big Splat has undoubtedly been easier for scientists to swallow because it assimilates the best points of all three classical theories while patching up their most obvious flaws. This has led Michael Drake to describe it as an "integrative" theory, a true synthesis of the things we have learned about the Moon.

For example, the giant impact theory resembles Darwin's fission hypothesis in fixing Earth as the "birth site" of the Moon. In addition, some of the material from the Moon still comes from Earth's mantle (although most comes from the impactor). But the impact hypothesis provides the main ingredient that the fission theory lacked: a huge energy and momentum source for ripping that material away from Earth. In addition, its random nature explains why Earth alone ended up with a large Moon, while Venus and Mars didn't.

As for See's hypothesis, the giant impact really is a capture of sorts—a capture gone awry, where the quarry got a little too close to the hunter. But at the same time, it is a great improvement on the capture theory in two respects. First, an impact is actually more plausible than a near miss from the point of view of celestial dynamics, because the focusing effect of gravity makes Earth a target that "wants" to be hit. (Think again of Comet Shoemaker-Levy and Jupiter.) Second, if you want to think of the impact as a sort of capture, it provides exactly what capture theorists such as See were missing: a resisting medium to slow the interloper down. What medium could be more resisting than solid Earth?

The giant impact also is a sort of coaccretion. The impactor and Earth do, after all, grow up by accretion in the same "feeding zone." The giant impact also leads inevitably to a (brief) stage where Earth is surrounded by a disk of material, much like the planetesimal swarm that the coaccretionists believed in. But it answers questions that the coaccretion theory couldn't. It explains why Earth and the Moon have so much angular momentum (the glancing blow from the impactor gave it to them). And it explains why the Moon accreted on such a rapid time scale: The ring was not a stable, long-lived phenomenon like Saturn's rings. It was a very crowded ring, with lots of debris. It was located outside the Roche limit, where the debris could stick together without being tidally disrupted, but at the same time it was not too far outside the Roche limit, where the material would have become more dispersed and accreted more slowly.

In short, the giant impact theory is a wonderful compromise. But Alastair Cameron, its coinventor, bristles if you talk about it that way. "It has something in common with all three theories, but that had nothing to do with the way it was developed," he says. "It is only superficially integrative in retrospect." Indeed, it was born out of a central realization that began with Baldwin and Shoemaker, grew bolder with the work of Safronov, and finally reached its logical culmination with Hartmann and Cameron: that *impacts* are the primary creative (and destructive) process in the solar system. The giant impact hypothesis never could have been designed by a committee attempting to reconcile the three classical theories. It only could have been devised by innovators who deliberately rejected the long-standing scientific prejudice against one-time catastrophes. So perhaps the best way to describe it is neither as pure synthesis or pure revolution, but as a little of both: a synthesis that was made possible by a revolutionary insight.

12

Introducing Theia

"The impactor." "The planet that is no more." These seem like feeble and inappropriate ways to describe a body that has had a more profound effect on the history of our planet than any other object in the Solar system—including the meteorite that killed the dinosaurs. This planet was the original, the true Atlantis: an entire world that vanished without a trace (well, almost) in a calamity of unimaginable proportions.

Nevertheless, for sixteen years after the Kona consensus and for twenty-six years after Bill Hartmann first proposed the giant impact theory, the star of the story went unnamed. In 2000, that sad situation was finally rectified by Alex Halliday, the same scientist who has provided us a first glimpse of that planet's chemistry through the obscuring haze of the Moon's violent history and the chaos of the giant impact itself. Halliday has begun calling her Theia—in Greek mythology, the mother of Selene, the Moon. His terminology has not become standard yet—other scientists still say "the impactor"— but it is an inspired choice, and I hope it will become the accepted name. More than that, I hope that nonscientists will learn to appreciate Theia as well. She is a large part of the Moon and a small part of Earth. She is the sacrifice that perhaps had to be made so that life as we know it could evolve on Earth.

"But Where's the Hole?"

One of the most puzzling things about the giant impact hypothesis, to me, is the fact that it remained virtually unknown to the public for nearly a decade and a half after the Kona conference. Shouldn't people know about the most Earth-shaking event in history? Shouldn't the giant impact be at least as famous as the extinction of the dinosaurs? And from a purely economic viewpoint, shouldn't taxpayers find out about the scientific payoff from their $25 billion investment in the Apollo missions?

Yet the Big Splat flew stubbornly under the popular radar for years. For example, compare the *New York Times*'s coverage of the twenty-fifth anniversary of the *Apollo 11* landing to its coverage of the thirtieth anniversary. In 1994, science columnist John Noble Wilford wrote a melancholy piece titled "25 Years Later, Moon Race in Eclipse." Wilford talked about the "unfulfilled promise" of the Moon missions, the Cold War politics that motivated the Apollo program, the lack of vision, and the loss of confidence that brought it to an end. Yet he did not write a single word about the scientific discoveries that had come out of the missions. The words "giant impact" were mentioned nowhere, even though this was ten years after Kona.

The thirtieth-anniversary coverage was as different as night and day. Under the headline "Apollo Opened Window on Moon's Violent Birth," William J. Broad wrote in glowing terms about the giant impact theory and its greater implications for all of science. "Scientists and historians now say that the lunar landings helped reveal the secret of cosmic violence, opening a line of inquiry still advancing today," Broad wrote. No more hand-wringing over what we went to the Moon for, or why we gave up so easily. For the first time, America's "newspaper of record" was telling the public that the Apollo landings were, scientifically, an outstanding success.

I can only guess why it took so long for the Big Splat to get to this point. Perhaps it was the unusual history of the theory. Unlike Luis and Walter Alvarez's theory of the meteorite that killed the dinosaurs, the giant impact theory had a long period of dormancy and then leaped overnight to orthodoxy. It skipped over the phase of being controversial, which might have attracted the attention of journalists. (This is, of course, a great oversimplification, because, as I have pointed out, some aspects of the giant impact *are* controversial.) Also, the giant impact didn't have a photogenic victim like the dinosaurs. However, that, too, may be a misperception. *Life itself* may have been a hostage to the Big Splat and the late heavy bombardment: some scientists have suggested that life may have tried to get started more than once, only to be killed by cataclysms and magma oceans.

At last, though, the Big Splat is making the transition from professional journals to popular science. My favorite example of this came when I visited Tucson, Arizona, to interview Al Cameron and Bill

Hartmann for this book. On an off day, I checked out the Arizona-Sonora Desert Museum and was thrilled to find an exhibit on geological history that mentioned the giant impact. A small girl, perhaps six years old, was behind me in line with her father. He read the caption to her, explaining how a Mars-size planet ran into Earth and exploded. With the unerring instinct children have for the really important questions, she asked him: "But where's the hole?"

If that girl is reading this book, here's your answer: there is no hole. The explosion didn't merely put a dent in Earth, it rearranged it completely. Earth blew up into an amoeba shape, and then settled back down into a sphere. The remnants of the great inferno have long since been buried underground on Earth or on the Moon. The clues are easy to find on the Moon, because it has remained undisturbed for billions of years. They are much better hidden on Earth, which destroys its own history through erosion and plate tectonics. But we will keep looking, with hammers or with spaceships or simply with our brains. Perhaps you will join the search someday.

The Big Splat (Condensed Version)

In the preceding chapters I dissected the various theories of the Moon's origin, including the giant impact theory, piece by piece. It is important to realize that many of these details are still uncertain; but at the same time, all the caveats and alternative versions have the unfortunate effect of obscuring the "big picture."

Sometimes, after immersing myself in the technical papers of scientists, I begin to lose my sense of reality. I forget that hafnium is a toxic gray element, that 2,200 degrees Kelvin would cook me so fast I wouldn't even be able to finish this sentence, and that 270,000 atmospheres of pressure would squash me flatter than a pancake even faster than that. I can't even begin to grasp how long 4.5 billion years is. Even the Moon becomes almost a figment of the imagination, rather than the reassuring silvery disk shining outside my window right now. Such is the power of modern science that it makes the familiar seem fabulous, and the fabulous seem ordinary.

To bring some order and restore a sense of reality to this jumble, I would like to end by recapitulating the whole Big Splat and its aftermath, leaving out the caveats. I believe that this timeline more or less reflects the consensus of lunar scientists today, although some details (particularly on the timing of the impact) are still being debated and some will undoubtedly turn out to be wrong.

Time Zero (4.57 billion years ago)   The solar nebula, a giant cloud of gas and dust, begins to collapse—perhaps triggered by the shock wave from a nearby supernova. The inner part becomes dense enough to initiate nuclear fusion, and the Sun is born. Within 1 or 2 million years, most of the gas has been blown away by the solar wind, and the solid material has condensed into a disk of rock and ice fragments, or "planetesimals."

Time Zero + 25 million years   Earth has already grown to 80 percent of its present size through the accretion of planetesimals. Its "feeding zone" is still chock full of small rocks and medium-size planetesimals. And Earth also has one full-fledged planet as a "roommate," Theia, which is about a tenth the size of Earth. Perhaps because it is smaller, Theia has already formed an iron core, while Earth has not.

Time Zero + 50 million years   Earth reaches 90 percent of its present size. Core formation has begun in earnest, perhaps abetted by planetesimal impacts. The feeding zone is clearing out now, and the impacts are becoming less frequent but more violent. Some of these may even be powerful enough to melt parts of Earth's surface, creating temporary magma oceans.

At this point, something happens to put Theia and Earth on a collision course—or perhaps Theia's luck just runs out. In the final days, the livid face of Earth grows larger and larger in Theia's sky; in the final minutes it fills the entire sky and blots out the Sun. And then . . .

Impact!   Theia plows into Earth with the energy of trillions of hydrogen bombs. Within hours, the former planet has been turned inside out. Its core is raining down on Earth in an iron deluge, while its lighter mantle has been stretched out into a long arm reaching thousands of miles into space. The misshapen Earth is spilling its guts like a watermelon shot with a rifle. Its suddenly lopsided gravitational field tugs the spiral arm of debris into orbit. Some of the debris also may get a boost from the tremendous explosion of gases produced by the impact.

After a week, most of the debris from the former Theia has rained back to Earth, but a substantial fraction—more than twice the mass of our present-day Moon—remains in orbit in a ring around Earth, possibly with several spokes or spiral arms. The spokes actually help pump material beyond the "Roche limit," roughly twelve thousand miles away, where the solid fragments can begin clumping together again. The temperature of Earth's surface is still several thousand degrees and the ring likewise, although it is cooling fast.

Impact + 1 year   Every night Earth's sky is lit up by fireballs descending from the heavens, and by the sparkles of light from rocks in the ring of debris colliding with each other. Once the ring has cooled down enough to form solid rocks, the clumps of rock grow at an astounding pace. The largest clump grows visibly from one week to the next, and begins to emit a ruddy glow from the heat of all the rocks raining down on it. Within a year, Earth has a new companion, a fiery red globe that people would one day name the Moon. At the moment, it looks more than ten times larger than the Moon of today, because it is that much closer.

Impact + 50 million years   The rain of debris on Earth is essentially over, and accretion is complete. The ocean of magma that covered Earth after the impact has solidified. Earth's core—which contains most of the iron from Theia as well—has finished separating from the mantle. The energy released by core formation causes many of the gases trapped inside Earth, including water vapor, to escape to the surface.

Meanwhile, the Moon's magma ocean also has solidified and a light, dazzlingly white crust of anorthosite has formed. It still looks like a fresh snowfall; there are no dark "seas" yet on the Moon's surface.

Impact + 100 million years   Earth's surface has cooled enough to allow water to condense. In a deluge that would have made Noah proud, the clouds of water vapor have rained out and formed oceans of water. Now the geology of the Moon and Earth start to diverge. On Earth, water shapes the continents, while underneath the surface it lubricates the mantle. The constant convection in Earth's mantle drives the motion of continental plates, which ultimately destroys all record of Earth's original crust. Meanwhile, a thick atmosphere—possibly of carbon dioxide at this point—causes smaller meteorites to burn up, so that only larger ones form craters.

On the Moon, it is just the opposite. The crust forms a solid shell, and there are no plate tectonics. But the constant pummeling of meteorites is putting pockmarks on the pristine face of the Moon and churning up its surface into powdery "regolith."

Impact + 600 million years (4 billion years ago)   In a late coda to the solar system's formation, a new deluge of large meteorites begin bombarding both Earth and the Moon. These meteorites come from the asteroid belt, from which they were dislodged by (take your pick) the last wild ride of Planet V, the excellent adventure of Uranus and Neptune as they migrated outward beyond Saturn ... or perhaps something else. On the Moon, the bombardment leaves huge scars that would form the major basins—Imbrium, Orientale, and others. On Earth, the barrage may have been equivalent to a dinosaur-killing asteroid every twenty years. It may have killed off the earliest life on the surface; on the other hand, it may have created an evolutionary niche for the thermophilic (heat-loving) life forms that seem to be the most primitive organisms on our present Earth.

3.8 billion years ago   The late heavy bombardment stops. Life gets a green light on Earth, although the precise time when it emerges is still unknown. The Moon is moving rapidly away from Earth, due to tidal friction, and slowing down Earth's rotation. It also makes Earth's axis of revolution more stable, so the poles do not precess as far as they would have if there had been no Moon. In future years this will make Earth's climate more temperate and hospitable to life.

3.6 billion years ago   The Moon's last geological moment of glory begins. The mare lavas, heated by the decay of radioactive uranium and thorium, melt and erupt to the surface, where they pool in the basins that have already been created by impacts. This episode of volcanism is limited almost exclusively to the near side. On the far side, where the crust is thicker, the lavas only manage to punch through in a couple places. Around the edges of the basins, "fire fountains" erupt in a showy spectacle, through fractures created by the subsidence of the basins.

On Earth there is enough heat from radioactive decay to keep volcanism going to the present day. On the Moon, however, the "heat engine" eventually runs out of fuel. The lavas stop flowing sometime after 3.2 billion years ago, the age of the youngest basalts collected by Apollo.

3.2 billion years ago (?) to the present   The Moon settles in for a long period of geologic quiet. If any volcanic activity is going on at all, it is very sporadic and very deep. The Moon does continue to have "moonquakes," which are tiny by Earth standards and have no effect at the surface; they are due to the ongoing tidal forces from Earth.

The face of the Moon continues to change, but only gradually, due to meteorite impacts. Occasionally a big impact, such as Tycho 109 million years ago, ejects bright streaks of debris all the way across the Moon. Some impacts are powerful enough to launch small rocks into space, and some of these rocks are captured by Earth's gravity. They fall back to Earth as meteorites, finishing a round trip that began 4.5 billion years ago.

While the Moon looks serenely on, almost unchanging, Earth goes through nearly its entire history. And what a history it is! Continents form and collide, separate and rejoin. Sea levels go up and down. For billions of years only one-celled organisms live on Earth— and then suddenly, in the blink of an eye, multicelled organisms grow and proliferate. One by one, plants, arthropods, fishes, dinosaurs appear. Glaciers march down from the polar regions, then retreat again. Now and then large meteorites still strike Earth. On the Moon such impacts barely disturb the eternal peace, but on Earth they create ecological holocausts and drive most of the species to extinction. But life always comes back with new species and new capabilities: animals that can fly, animals that can think.

About 30 years ago, a few Moon rocks find a new route back to Earth, in the storage bay of spaceships built by these thinking creatures. And a few Earth items make the reverse journey—an ungainly, spiderlike lander, a hammer and a feather, a plaque that reads "We Came in Peace for All Mankind." This new "geological process" seems to have stopped for the time being, but it could begin again, whenever the thinking life forms on Earth decide it is worth their trouble.

Whenever I see a timeline like this, I always wonder: What comes next? In this case, I think I know. The Moon's history cannot be over. It will be reclaimed by those thinking creatures, whether it takes a hundred years, a thousand, or a million. For the Moon, a million years is not a long time. The Moon is a survivor, just like us.

Appendix

Did We Really Go to the Moon?

Bill Hartmann, the astronomer/artist/science fiction writer who has played such a large role in this story, once wrote a story of his own for *Nature*, a science magazine that normally prints only nonfiction. In the story, called "The Paradigm and the Pendulum," a boy named David, living in the year 2063, explains in an essay why the so-called exploration of space never happened. Even though his own grandfather had flown to the asteroid belt, David (and most other Americans) had been brainwashed by a fundamentalist religious movement into believing that his grandfather's stories were a lie.

Amazingly, the reeducation that Hartmann predicted has already begun. In February 2001, a year after his story appeared and sixty-two years before his predicted dystopia, the Fox television network twice aired a documentary called "Conspiracy Theory: Did We Land on the Moon?" Fox interviewed several self-appointed experts who claim that the Apollo landings were faked, either to fool the Soviets about America's space capabilities or to avoid the public-relations debacle of failing to reach the Moon before President Kennedy's 1970 deadline. Hartmann says that he has gotten many e-mails since the Fox special ran, from well-intentioned people asking if the Apollo landings really were a hoax. "This is a very strange thing," he says, "to live through something and participate in it and then, in your older age, to see people getting on television and saying it never happened."

One of the conspiracy theorists, a filmmaker named Bart Sibrel, has a web site, *http://www.moonmovie.com*, devoted to marketing his own video, "A Funny Thing Happened on the Way to the Moon." On the web site he lists his "Top Ten Reasons Why No Man Has Ever Set

Foot on the Moon." Although some standard arguments of Moon hoaxers do not make it into Sibrel's top ten, it is a good starting place for understanding their arguments. Here are Sibrel's ten reasons, and ten possible responses.

- *Sibrel:* "10. 'Tricky Dick' Richard Nixon was president at the time. He was the king of cover-up, secret tapes and scandal."

  Response: If the Moon landings were faked, the wheels had to be set in motion long before Nixon became president. Sibrel's other arguments (particularly no. 7) imply that *Apollo 8*, the first manned circumlunar mission, would have had to be faked, too—and this mission flew while Lyndon Johnson was still in the White House. You can't pin this one on Tricky Dick!

- *Sibrel:* "9. A successful manned mission to the moon offered a wonderful pride-boosting distraction for the near revolt of the citizens of America over 50,000 deaths in the Vietnam War."

  Response: This is a motive, not a proof of guilt. It remains true even if the Moon landings were not faked.

- *Sibrel:* "8. The Soviets had a five-to-one superiority to the U.S. in manned hours in space. They were first in achieving the following seven important milestones."

  Response: Since the collapse of the Soviet Union, the reasons why the Soviets failed to place a man on the Moon have been well documented. Most important was the fact that their giant N-1 booster, the equivalent of the Saturn 5, exploded twice on the launch pad. Also, the death in 1966 of Sergei Korolev, their chief rocket designer, left a leadership void that was never truly filled. Finally, Soviet electronic and computer technology was far behind America's, not only in the 1960s but throughout the Soviet era. It is remarkable enough that Americans made it to the Moon with their 1960s-era computers; it would have been even more difficult for the Soviets.

- *Sibrel:* "7. Passengers of a spacecraft that went further than Earth orbit would likely have been subject to lethal radiation." (He explains that he is referring to the Van Allen radiation belts.)

  Response: It's true that astronauts were exposed to radiation, and the Van Allen belts are not a place where you would want to spend a vacation. But the astronauts passed through the most intense radiation in about five minutes. Professionals estimate that the dose of radiation they received was about 2 rems. A lethal dose is considered to be 250 rems. The risk was not negligible, but Armstrong and Aldrin were military pilots who courageously volunteered for a mission they knew to be hazardous. This brings us to Sibrel's next argument:

- *Sibrel:* "6. Neil Armstrong, the first man to supposedly walk on the moon, refuses to give interviews to anyone on the subject. . . . Collins also refuses to be interviewed. Aldrin, who granted an interview, threatened to sue us if we showed it to anyone." (Sibrel claims that they won't talk because they don't want to lie.)

Response: All three astronauts did give a press conference on August 12, 1969. They even addressed a point that conspiracy theorists are still harping on years later. The Fox TV documentary suggested that the *Apollo 11* photographs are fake because there are no stars in the background. Yet Armstrong had stated in 1969, "We were never able to see stars from the lunar surface or on the daylight side of the Moon by eye, without looking through the optics." The conspiracy theorists forget that Armstrong and Aldrin landed during *daytime;* in fact, it was lunar morning throughout their mission. Though there was no atmosphere and the sky was black, the glare from the Moon's surface was extremely bright, and their pupils, adapting to it, would not have been able to see the faint stars. The same thing goes for the cameras they used to take pictures on the Moon.

As for being reluctant to talk and not wanting to lie, there is certainly nothing in the press conference transcript to suggest this. In fact, there is even some sly humor. When one reporter asked if the astronauts had ever been "spellbound" during their walk on the Moon's surface, Armstrong said, "About two and a half hours." Armstrong also mentioned a prank they pulled on Mission Control on the way home, playing weird space sounds from a tape recorder over the intercom. Does this sound like someone who is lying under duress?

As for Sibrel's problems with Aldrin, consider this scenario. You have spent six years of your life preparing for a dangerous mission that no one in history has ever done before. At the end of that time, you pull it off flawlessly. Thirty years later, a filmmaker shows up at your door saying that it was all a fraud. Do you think that you might get a little testy?

- *Sibrel:* "5. Newly retouched photographs correct errors from previously released versions. Why would they be updating thirty-year-old pictures if they really went to the moon?" (Sibrel shows an example where a shadow in the shape of a "C" has been removed; he calls the "C" a "prop ID.")

  Response: The idea that the single letter "C" is a prop identification is somewhat amusing; a style more typical of NASA would be a seven-digit control number. As for the first question: Why not? Why do movie buffs restore old movies and car buffs restore old cars to pristine condition? They want to see things the way they really were, without all the scratches.

- *Sibrel:* "4. Rediscovered lost footage shows the American flag blowing in the wind." (This is supposed to show that the Moon landing was faked, because the Moon has no atmosphere and therefore no wind.)

  Response: The unfurling of the flag caused it to vibrate, and with no atmosphere to dampen the vibrations they could have gone on for some time.

- *Sibrel:* "3. Enlarged photographs underneath the lunar lander's 10,000 lb. thrust engine show the soil completely undisturbed."

  Response: The idea that there would have been a crater underneath the lander seems to have originated with another conspiracy theorist, Bill Kaysing, who used to work for a company that helped construct the Moon landers. It has the sound of an engineering concern that was brought up before the missions, and has now been revived after the fact as a "proof" that the landings were faked. In any event, the astronauts throttled back the engines to much less than full power before landing. Moreover, there were six-foot-long sensors on the footpads, and the astronauts were supposed to cut the engines as soon as the sensors contacted the ground. More than one crew commented on the rather rough landings that ensued. Finally, it's not clear that the photographs show the soil "completely undisturbed," as Sibrel says—in fact,

some of the dust seems to have blown away, and some anticonspiracy theorists say they can see scorch marks.

- *Sibrel:* "2. Rare, uncirculated photographs, allegedly from the
  moon's surface, show scenes supposedly lit solely by sunlight. Yet
  they contain shadows that do not run parallel with each other,
  indicating supplemental artificial light."

  Response: "Rare" and "uncirculated" are inflammatory words and
  not true anymore: The photographs are all over the Internet now!
  It is hard to believe that anyone could be bothered by some of
  them. One picture, taken by remote control, shows an astronaut
  who is clearly standing on a rise and whose shadow runs down
  the slope. The other astronaut is *not* on a rise and—big surprise!—
  his shadow points in a different direction and is not as long. It is
  also worth pointing out that the astronauts all commented on
  the difficulty of gauging distances and slopes on the Moon; you
  could not even tell if you were standing upright or leaning a bit.
  Often the scientists watching in Houston would ask the astronauts to sample a rock that appeared nearby, only to be told that
  it was out of walking distance. Given the strange, literally
  unearthly lighting conditions, it is not surprising that a conspiracy theorist specifically hunting for optical illusions would be
  able to find some photographs that do not look "right."

- *Sibrel:* "1. Recently uncovered mislabeled, unedited, behind-the-
  scenes video footage, dated by NASA three days after they left for
  the moon, shows the crew of *Apollo 11* staging part of their pho-
  tography."

  Response: This is Sibrel's big find, and the one he wants you to
  pay him to see. I will yield the floor to an anticonspiracy theorist
  on the Web who writes, "Others who have seen this film (and
  much regret wasting their money) are more of the opinion that it
  simply shows the crew preparing camera angles and lighting
  before a live TV interview."

Now I will offer one and only one reason we can be sure that men
did land on the Moon. It's 841 pounds of reasons: the lunar rocks
and soil samples. These are conspicuously absent from Sibrel's list.
Sibrel says that he has enough proof of the alleged Moon hoax

to stand up in a court of law, but I cannot imagine that any court would accept debatable interpretations of photographs over material evidence.

Could the rocks have been faked? In a word, no. The *Apollo 11* samples went out to 142 qualified investigators in countries around the world, from Switzerland to Japan. The investigators in foreign countries had no reason to cooperate with a hoax, and would surely resent any implication that they were tools of the American space program. The great majority of the U.S. investigators were *not* employees of NASA or the federal government but professors at universities, and they would have *everything to lose* by participating, wittingly or unwittingly, in a sham. None of these scientists reported any concerns about the authenticity of the rocks.

Not only that, at the time of the Apollo missions, the United States had a rival that surely would have been eager to expose anything that looked like a hoax. Yet, to my knowledge, the Soviets never questioned whether Americans had actually landed on the Moon.

Moving from political psychology to science: the Apollo rocks were formed under conditions that could not be duplicated in any laboratory, no matter how lavishly funded. The breccias were fused by impacts that were billions of times more energetic than any nuclear explosions. Most Moon rocks are older than any Earth rocks, and as Randy Korotev (see below) says, "Anyone who figures out how to fake that is worthy of a Nobel prize." The Moon hoaxers are giving humans too much credit, and not showing enough awe before the powerful forces of nature. There are plenty of things that nature can do that man cannot duplicate.

Furthermore, if the Moon rocks had been faked, they surely would have been faked using pre-Apollo guesses of what to expect. They would probably have been Ureyan cold-Moon rocks, or else volcanic rocks less than a billion years old as the "hot-Mooners" expected. They would not have contained so many surprises, such as the green glass. It is odd, too, if the Moon rocks were faked, that they so perfectly matched a class of meteorites that had *not been discovered yet*, the two dozen or so meteorites that are believed to have come from the Moon. The first of these was discovered in 1979; most of them were discovered in Antarctica beginning in the 1980s. Did NASA, perhaps, send crews to Antarctica to plant bogus meteorites? I suppose that a conspiracy theorist would say so.

Again, I would like to emphasize that the scientists who have studied the Moon rocks and the Moon meteorites are not nameless, *X-Files*-ish "government scientists." They work at private or public universities and can easily be contacted by telephone or by e-mail. I will leave the last word with one of them, Randy Korotev of Washington University in St. Louis, who participated in some of the Antarctic meteorite-hunting expeditions and wrote on his web site:

"Any geoscientist (and there have been thousands from all over the world) who has studied lunar rocks knows that anyone who thinks the Apollo lunar samples were created on Earth as part of a government conspiracy doesn't know much about rocks. The Apollo samples are just too good. They tell a self-consistent story with a complexly interwoven plot that's better than any story any conspirator could have conceived. I've studied lunar rocks and soils for thirty-plus years and I couldn't make even a poor imitation of a lunar breccia, lunar soil, or a mare basalt in the lab. And with all due respect to my clever colleagues in government labs, no one in 'the government' could do it either, even now that we know what lunar rocks are like."

Glossary

accretion disk   A wide and thin disk of dust and ice that forms around the Sun (or around any star) early in the process of planetary formation.

angular momentum   A quantity that measures how hard an object is spinning. The law of conservation of angular momentum says that this quantity can only be changed by forces from outside the spinning object. Thus, for example, the combined angular momentum of Earth and the Moon has hardly changed in the past four billion years.

anorthosite   A rock with a greater than 90 percent concentration of plagioclase feldspar. Because such rocks are light, they would have floated to the top of a magma ocean. Thus they represent pieces of the Moon's original crust.

apogee   The greatest distance between the Moon (or any Earth satellite) and Earth.

basalt   A dark, dense rock type formed by solidifying lava. On the Moon, such rocks are mostly found in the maria, thus giving rise to the phrase "mare basalt."

breccia   A composite rock consisting of fragments that have been welded together, usually by the shock of a high-speed impact. Uncommon on Earth but extremely common on the Moon.

capture hypothesis   The hypothesis that the Moon formed separately from Earth, in a different part of the solar system, and was then captured after passing very close to Earth.

center of mass   A point within an object or system of objects that represents their average position. In many physical problems the motion of the center of mass is much simpler than the motions of the objects themselves. For example, it is not technically correct to say that the Moon orbits around Earth; instead, they *both* orbit around the center of mass of the Earth-Moon system, which is about a thousand miles underneath Earth's surface.

centrifugal force   The apparent outward force felt by an object in circular motion—for example, the occupants of a car turning a corner. It is sometimes called a "fictitious force" by physicists because it is actually due to the acceleration of the object's frame of reference. Nevertheless, it remains a handy and easily understandable concept.

centripetal force   The inward force that causes an object to remain in circular motion.

coaccretion hypothesis   The theory that the Moon formed by accretion of planetesimals within Earth's feeding zone and was never an independent planet.

core   The innermost part of a planet. Earth's core is believed to be mostly molten iron and nickel, and it comprises 32 percent of our planet's mass. The Moon's core, by contrast, contains only 2 to 4 percent of its mass. This discrepancy is an important constraint on all theories of the Moon's origin.

craters   The dominant geological features on the Moon. For a while a major controversy centered on whether the craters were produced by volcanic activity or by meteorite impacts; now all lunar scientists agree that the vast majority are impact craters. Not a single crater on the Moon has been definitively proven to be volcanic.

deuterium   A heavy isotope of hydrogen whose nucleus contains one proton and one neutron instead of a lone proton.

ecliptic plane   The plane in which Earth orbits around the Sun. The other planets, except Pluto, also orbit in nearly the same plane.

ellipse   An oval shape that can be most simply described as a circle seen from an oblique angle. An ellipse has two focal points or "foci," and the sum of the distances from any point on the ellipse to the two foci is a constant. According to Kepler's first law, the orbit of a planet about the Sun is an ellipse, with the Sun at one focus (not at the center).

ellipsoid   A solid object whose cross sections are all ellipses. An ellipsoid has three axes, which may have three different lengths. If they all have the same length, the ellipsoid is a sphere. If two of them have the same length, it is sometimes called a spheroid. The Moon is a spheroid: its front-to-back diameter is practically identical to its east-to-west diameter, but its north-to-south diameter is slightly less.

escape velocity   The upward velocity that is enough to enable an object to escape a planet's gravitational field. On Earth, escape velocity is about twenty-five thousand miles per hour.

feeding zone   The band centered on the Sun from which an accreting planet draws the great majority of its material, in the theory of Safronov.

fission hypothesis   The hypothesis that the Moon was formed when the rapidly spinning, molten Earth flung off a large chunk of itself into orbit.

fusion   A process by which two atomic nuclei are joined to form a heavier nucleus, releasing energy in the process. The Sun is powered by the fusion of hydrogen and deuterium atoms to form helium.

giant impact hypothesis ("Big Whack," "Big Splat")   The hypothesis that the Moon was formed by a collision between Earth and another planet. "Big Whack" is an informal name that has occasionally been used for this hypothesis; "Big Splat" is another informal name, coined for this book.

glass   A disordered, noncrystalline material formed by rapid cooling of a silicate melt. Much of the Moon's "soil" is actually ground-up glass, created when lava vented into the extreme cold of space.

gravity   One of the four fundamental forces in the universe known to physicists. It pulls any two massive objects together; however, it is so weak that it only becomes significant on planetary scales. Newton's law of gravitation describes it but does not explain why it exists. See *also* Newton's law of gravitation.

heliocentric theory   A theory, such as the ones proposed by Aristarchus and Copernicus, that places the Sun at the center of the universe. It is actually more correct, if pedantic, to call Copernicus's theory *heliostatic*—that is, the Sun stands still. Already in Copernicus's time it was evident from observations that the Sun was not literally at the center of Earth's orbit.

highlands   Generally speaking, any terrain on the Moon that is not covered by mare basalt. Ironically, the Apollo missions discovered that the mare basalts are not very thick, and underneath them is more "highland"-type rock.

ilmenite   A dark and heavy iron titanate found particularly in the Mare Tranquillitatis basalts and that is probably a significant constituent of the lunar mantle.

Imbrian Basin   The largest basin on the Moon's near side. Contains the Imbrian Sea (Mare Imbrium).

impact basin   A basin, such as Imbrium or South Pole-Aitken, that has been created by an impact blast. On the near side of the Moon, the main basins have been filled by lava to create maria, or "seas"—the dark spots that make up the "man in the Moon." However, a basin and a sea are not necessarily the same thing: the South Pole-Aitken basin on the far side, the Moon's largest and deepest, has not been filled in at all. Basins are distinguished from craters primarily by their size; a single basin may contain hundreds of smaller craters.

impact melt   A small, glassy region in a rock, or sometimes a small pool of glass at the center of a crater, that indicates the rock was melted by a meteorite impact.

isostatic adjustment   The tendency of a crater floor to rebound (over a long period of time) because of the weight that has been removed from it.

lithophile ("rock-loving") element   An element that enters into rock more readily than it alloys with iron.

luminiferous aether   A substance that was theorized, in the nineteenth century, to be the medium through which light propagates. A famous experiment by Michelson failed to detect it, and Einstein's theory of relativity essentially showed that it cannot exist.

magma ocean   An ocean of molten rock ("magma") that covers most or all of a planet's surface. The Moon almost certainly had a magma ocean after it formed, and the giant impact hypothesis seems to require that Earth had one, too.

mantle   The part of Earth lying between the crust and the core; sometimes divided into two zones, the "upper" and "lower" mantle. It consists primarily of solid rock.

many-moons theory   A briefly popular version of the capture hypothesis, according to which Earth captured several moonlets. These moved outward from Earth at different rates due to tidal friction and thus eventually collided with each other and formed our present Moon.

mare (plural: maria)   A dark, basalt-filled plain. Early Moon observers believed that these regions were filled with water, and hence applied the Latin word *mare* (sea) to them. The name stuck.

Mare Orientale   A spectacular "bull's-eye"-shaped sea on the Moon's far western limb.

Mare Tranquillitatis (Sea of Tranquility)   Site of the first lunar landing, by *Apollo 11*.

moment of inertia   A quantity that measures the resistance of an object to turning or to any changes in its rotational motion. When multiplied by the angular velocity, it gives the angular momentum.

month   Informally, the length of one lunar cycle. However, there are several different kinds of month. The *synodic* month measures the time from new Moon to new Moon, which averages about 29.53 days. The *sidereal* month measures the time it takes for the Moon to reach the same position against the background of stars, and averages 27.32 days.

nebular hypothesis   The theory, originally due to Kant and Laplace, that the solar system formed by the cooling and contraction of a giant cloud of gas (or "nebula").

Newton's law of gravitation   A quantitative description of the gravitational attraction between any two objects. According to Newton's law, the attraction is proportional to the size of the two objects. It drops off very rapidly as the objects move apart.

Oceanus Procellarum (Ocean of Storms)   The Moon's only official "ocean," this is the largest and most ragged dark spot on the near side. It could be a very weathered impact basin, akin to South Pole-Aitken on the far side, but lunar scientists are not sure at present.

olivine   An iron and magnesium silicate, this is one of the most common minerals in both Earth's and the Moon's mantles.

perigee The closest distance between the Moon (or any Earth satellite) and Earth.

perihelion The closest distance between Earth (or any solar satellite) and the Sun. At present, Earth reaches its perihelion in the first week of January, despite the common misconception that Earth is farthest from the Sun in winter. In fact, the opposite is true (at least in the Northern Hemisphere)!

plagioclase feldspar An aluminum silicate, this is one of the most common lunar minerals, and is particularly abundant in the highlands.

planetesimal hypothesis The theory, proposed by Chamberlin and Moulton, that the solar system formed by the accretion of small rocks ("planetesimals"). Though this was conceived as an alternative to the nebular hypothesis, modern planetary scientists accept some elements of both theories.

precession A slow variation in the axis of rotation of any spinning object, such as a top. Earth's axis precesses over a period of 25,800 years. The motion of the line of nodes (the intersection of the Moon's orbital plane with the ecliptic) is also called precession; it takes 18.6 years to turn in a full circle. This gives rise to an 18.0-year periodicity in solar and lunar eclipses, called the Saros cycle, which has been known since ancient times.

prograde A rotation or orbital motion that is in the same direction (west-to-east) as that of most objects in the solar system.

proto-Earth (proto-Sun, etc.) A term sometimes used for Earth before it had finished accreting (or, in Laplace's nebular theory, contracting).

pyroxene An iron and magnesium silicate (with a slightly different formula from olivine), this is another common mineral in Earth's and the Moon's mantles.

quintessence A fifth element (different from earth, air, fire, and water) that Aristotle conjectured to be the building material for all heavenly bodies. Galileo's observations of the Moon, which showed mountains and Earth-like terrain, refuted Aristotle's centuries-old theory.

refractory element An element with a high boiling point.

regolith A layer of crushed rock, churned up by aeons of meteorite impacts, at the Moon's surface. This is sometimes referred to (especially in popular accounts) as the "lunar soil," a name that is rather misleading because it contains no organic material, no water, and is mostly sand.

resonance A common physical phenomenon in which oscillatory motions are greatly amplified if their frequency happens to coincide with the "natural frequency" of the oscillating object. This is, for example, how a trained singer can shatter a glass with his or her voice.

retrograde   An orbital or rotational motion that is opposite to the prevailing (west to east) motion of most objects in the solar system.

revolution   The motion of planets around the Sun, or of moons around planets. Because the Moon always shows the same face to Earth, its period of revolution is the same as its period of rotation: one month.

Roche limit   The distance (roughly 2.9 times Earth's radius) inside of which a moon cannot form because tidal forces would tear it apart.

rotation   The motion of a moon or a planet about its own axis.

runaway accretion   The stage of planetary accretion when the gravitational field of a growing planetesimal starts actively sucking in debris from the feeding zone. As a rule of thumb, this occurs when the planetesimal is roughly a thousand kilometers in diameter.

siderophile ("iron-loving") element   An element that alloys naturally with iron.

silicate   Literally, any compound containing silicon and oxygen. Because rocks are primarily composed of silicates, geochemists often use this as a fancy synonym for "rock."

South Pole-Aitken Basin   The Moon's largest basin, and the deepest excavation in the entire solar system. It is on the far side, and extends roughly from Aitken Crater to the South Pole. Some of the mountains at its edge can be seen on the near side near the South Pole.

terminator   The dividing line between the illuminated and dark portions of the Moon. The terminator moves as the Moon rotates; there is no permanent "dark side of the Moon."

tidal force   A force exerted by one large object (such as a planet) on another, nearby large object (such as a moon). The tidal force results from the difference between the gravitational attraction on the near side of the second object and the weaker attraction on the far side. This force tends to stretch the second object, and it can even tear that object apart if it is too close to the first. Tidal forces of the Moon on Earth, Earth on the Moon, and the Sun on both of them have played an extremely important role in the evolution of the Earth-Moon system.

Tycho   The brightest and one of the youngest craters on the Moon's near side, roughly 109 million years old. Named after the astronomer Tycho Brahe.

viscosity   A measure of how thick or resistant to flowing a fluid is. Molasses, for example, is more viscous than water.

volatile element   An element with a low boiling point.

vortex theory   A theory of Descartes that attributed the rotation and orbital motion of planets to vortices in space. This theory was very popular for about a hundred years, but eventually became obsolete because it was only a qualitative theory, while Newton's laws gave quantitative, testable (and correct!) predictions.

year   Informally, the amount of time it takes Earth to complete a revolution around the Sun. As with the month, there are different possible definitions of a year. The *tropical* year measures the time from one spring equinox to the next, and currently is about 365.24 days. The *sidereal* year measures the time it takes the Sun to reach the same position against the background of stars, and currently is about 365.26 days. The discrepancy is primarily due to the precession of Earth's axis.

zodiac   A band of constellations lying in or near the ecliptic plane. Ancient astronomers noticed that the planets and the Sun appear to follow the same path through this band of constellations, and this observation provided the factual foundation for the very speculative subject of astrology.

References

General Interest

For people who miss the days when "space exploration" meant going beyond low Earth orbit, there has recently been a deluge of good books about the Apollo missions, from nearly every conceivable point of view. Start with Andrew Chaikin's *A Man on the Moon*, the bible of Moon exploration, which inspired Tom Hanks's miniseries *From the Earth to the Moon*. (Or, if you prefer, buy the miniseries. It's on VHS and DVD.) For an astronaut's perspective, try *Apollo: An Eyewitness Account* by Alan Bean *(Apollo 12)*, *The Last Man on the Moon* by Eugene Cernan *(Apollo 17)*, or *Carrying the Fire* by Michael Collins *(Apollo 11)*. The view from the ground is told in Chris Kraft's *Flight: My Life in Mission Control*. I found *Exploring the Moon: The Apollo Expeditions* by David M. Harland, a British journalist and Moon junkie, to be very useful. With his blow-by-blow descriptions of what happened on the Moon, you'll swear he was right there looking over the astronauts' shoulders. A splendid book that gives the geologist's perspective is Don Wilhelms's *To a Rocky Moon*. For a book with fabulous pictures try *Full Moon* by Michael Light and Andrew Chaikin. If you like original press releases and documentation and other minutiae, the series of *NASA Mission Reports* published by Apogee Books makes for interesting browsing.

In the expository category, two up-to-date books I can recommend are *On the Moon* by Patrick Moore and *The Once and Future Moon* by Paul Spudis. *The Moon Book,* by Kim Long, is a handy reference book that covers everything from eclipses to librations in straightforward, no-nonsense fashion. If you want to observe the Moon through a telescope or binoculars (and want to know what you're looking at), procure a copy of *Atlas of the Moon* by Antonin Rukl and Thomas Rackham.

As far as I can tell there have been no book-length treatments of the giant impact hypothesis for a popular audience. (That is why I have written this book.) The closest thing is chapter 4 in Stephen G. Brush's *Fruitful Encounters*, vol. 3. (See my chapter 7 references.) At eighty-three pages it is a very long and substantive chapter. Brush delves even more deeply into the scientific nitty-gritty than I have, but his account is nevertheless very readable.

Another large category of Moon books deals with the folklore and mythology of the Moon. One of my favorites in this genre is simply called *The Moon*, by Maryam Sachs, a fascinating and beautifully printed smorgasbord of art, poetry, lists, and factoids. For readers who want to go beyond the smorgasbord approach I recommend *Women's Mysteries: Ancient and Modern* by M. Esther Harding. In spite of its title, Harding's book is mostly about the history of lunar mythology, and it lucidly explains why this mythology is still relevant today.

Finally, if you are an art lover, check out *Susan Seddon Boulet: A Retrospective* by Michael Babcock et al. Boulet was a popular West Coast artist who incorporated a moon—sometimes several of them—in nearly every canvas. "It's almost like a signature," she once said. "It felt that when I put the moon in, it was finished."

Chapter by Chapter

1: *A Highly Practical Stone*

Aveni, Anthony. *Stairways to the Stars*. New York: John Wiley & Sons, 1997.

Cinzano, Pierantonio, Fabio Falchi, and Chris Elvidge. "The First World Atlas of the Artificial Night Sky Brightness," *Monthly Notices of the Royal Astronomical Society* 328, no. 3 (2001): 689–707.

Courchesne, Luc. "Fragile Nights: A Collection of Ideas on Light, Darkness, and Human Behavior." Master's thesis, Massachusetts Institute of Technology, 1984. Reproduced at *http://www.din.umontreal.ca/courchesne/fragile .html*.

Hesiod. *Works and Days*. Translated by Hugh G. Evelyn-White. Cambridge: Harvard University Press, 1914. Reproduced at *http://www.perseus.tufts.edu*.

Plato. *Timaeus and Critias*. Translated by Desmond Lee. New York: Penguin, 1977.

Tøndering, Claus, et al. "Calendars through the Ages." At *http://webexhibits .org/calendars*.

Whitehouse, David. "Oldest Lunar Calendar Identified." BBC News, October 16, 2000. Reproduced at *http://news.bbc.co.uk/hi/english/sci/tech/news_id975000/975360.stm*.

Wolkomir, Joyce, and Richard. "'When bandogs howle & spirits walk': Studying the Nighttime Hours across the Centuries." *Smithsonian*, January 2001, 38.

2: *The Stone Star*

Aristotle. *On the Heavens*. Translated by W. K. C. Guthrie. Cambridge: Harvard University Press, 1960.

Burnet, John. *Early Greek Philosophy.* London: A & C Black, 1920. Reproduced at *http://plato.evansville.edu/public/burnet.* (This reference has the quote from Diogenes that provided the title for this chapter.)

Gottlieb, Anthony. *The Dream of Reason: A History of Western Philosophy from the Greeks to the Renaissance.* New York: W. W. Norton, 2000.

Hetherington, Barry. *A Chronicle of Pre-Telescopic Astronomy.* Chichester, Eng.: John Wiley & Sons, 1996.

Koestler, Arthur. *The Sleepwalkers.* New York: Macmillan, 1968.

Montgomery, Scott. *The Moon and the Western Imagination.* Tucson: University of Arizona Press, 1999.

Nicolson, Marjorie Hope. *Voyages to the Moon.* New York: Macmillan, 1948.

North, John. *Norton History of Astronomy and Cosmology.* New York: W. W. Norton, 1995.

Plutarch. "On the Face Which Appears in the Orb of the Moon." In *Moralia,* vol. XII, translated by Harold Cherniss. Cambridge: Harvard University Press, 1957.

———. *Plutarch's Lives.* Translated by Bernadette Perrin. Cambridge: Harvard University Press, 1914. Reproduced at the Perseus Digital Library, *http://www.perseus.tufts.edu.*

3: *Kepler Laughed*

Beer, A., and P. Beer, eds. *Kepler: Four Hundred Years.* Vol. 18 of *Vistas in Astronomy.* Oxford: Pergamon, 1975.

Casper, Max. *Johannes Kepler.* Stuttgart: W. Kohlhammer, 1948.

Drake, Stillman. *Galileo at Work: His Scientific Biography.* Chicago: University of Chicago Press, 1968.

Galilei, Galileo. *Sidereus Nuncius, or the Sidereal Messenger.* Translated by Albert van Helden. Chicago: University of Chicago Press, 1989.

Helden, Albert van. "The Galileo Project." At *http://es.rice.edu/ES/galileo.html.*

Kepler, Johannes. *Conversation with Galileo's Sidereal Messenger.* Translated by Edward Rosen. New York: Johnson Reprint, 1965.

Koestler, Arthur. *The Sleepwalkers.* New York: Macmillan, 1968.

Lear, John. *Kepler's Dream.* Translated by Patricia Frueh Kirkwood. Berkeley: University of California Press, 1965.

Montgomery, Scott. *The Moon and the Western Imagination.* Tucson: University of Arizona Press, 1999.

Nicolson, Marjorie Hope. *Voyages to the Moon.* New York: Macmillan, 1948.

Sobel, Dava. *Galileo's Daughter.* New York: Penguin, 2000.

4: *The Clockwork Solar System*

Beer, A., and P. Beer, eds. *Kepler: Four Hundred Years.* Vol. 18 of *Vistas in Astronomy.* Oxford: Pergamon, 1975.

Burton, David M. *The History of Mathematics: An Introduction.* Newton, Mass.: Allyn & Bacon, 1985.

David, F. N. "Some Notes on Laplace." In *Bernoulli, 1713. Laplace, 1813. Anniversary Volume,* edited by Jerzy Neyman and Lucien M. LeCam. New York: Springer-Verlag, 1965.

Davis, Charles Henry. "The Story Behind the Seal." In *Astronomical and Meteorological Observations Made at the U.S. Naval Observatory during the Year 1865.* Washington, D.C., 1867. Reproduced at *http://www.usno.navy.mil/USNOSeal.html.*

Diacu, F., and P. Holmes. *Celestial Encounters: The Origins of Chaos and Stability.* Princeton, N.J.: Princeton University Press, 1996.

Dick, Steven. *Plurality of Worlds: The Origins of the Extraterrestrial Life Debate from Democritus to Kant.* New York: Cambridge University Press, 1982.

Kant, Immanuel. *Universal Natural History and Theory of the Heavens.* Translated by Stanley L. Jaki. Edinburgh: Scottish Academic Press, 1981.

Moulton, Forest Ray. *An Introduction to Celestial Mechanics,* 2nd ed. New York: Macmillan, 1914.

Newton, Isaac. *The Principia: Mathematical Principles of Natural Philosophy.* Translated by I. Bernard Cohen and Anne Whitman, assisted by Julia Budenz. Berkeley: University of California Press, 1999.

Numbers, Ronald. *Creation by Natural Law: Laplace's Nebular Hypothesis in American Thought.* Seattle: University of Washington Press, 1997.

Serway, Raymond. *Physics for Scientists and Engineers,* 4th ed. Philadelphia: Saunders, 1995. (*Any* introductory physics textbook will do if you want to look up the basic material on centripetal acceleration, Kepler's laws, and Newton's law of universal gravitation.)

Sobel, Dava. *Longitude: The True Story of a Lone Genius Who Solved the Greatest Scientific Problem of His Time.* New York: Walker, 1995.

5: *Daughter Moon*

Brown, E. W. "The Scientific Work of Sir George Darwin." In George Darwin, *Scientific Papers,* vol. 5. Cambridge: Cambridge University Press, 1916.

Brush, Stephen G. "Early History of Selenogony." In *Origin of the Moon,* edited by W. K. Hartmann, R. J. Phillips, and G. J. Taylor. Houston: Lunar & Planetary Institute, 1986.

Chandrasekhar, Subrahmanyan. *Ellipsoidal Figures of Equilibrium.* New Haven, Conn.: Yale University Press, 1969.

Darwin, Charles. Letters. Charles Darwin Collections, Cambridge University Library, Cambridge.

Darwin, Francis. "Memoir of Sir George Darwin." In George Darwin, *Scientific Papers,* vol. 5. Cambridge: Cambridge University Press, 1916.

Darwin, George. Letters and reminiscences. Charles Darwin Collections, Cambridge University Library, Cambridge.

——— . *The Tides and Kindred Phenomena in the Solar System.* San Francisco: W. H. Freeman, 1962.

Goldreich, Peter. "Tides and the Earth-Moon System." *Scientific American,* April 1972, 42–52.

Keynes, Margaret. Reminiscences. Charles Darwin Collections, Cambridge University Library, Cambridge.

Lewis, Cherry. *The Dating Game: One Man's Search for the Age of the Earth.* Cambridge: Cambridge University Press, 2000.

Mitroff, Ian. *The Subjective Side of Science.* New York: American Elsevier, 1974.

Raverat, Gwen. *Period Piece.* New York: W. W. Norton, 1952.

Wood, John. "Moon over Mauna Loa: A Review of Hypotheses of Formation of Earth's Moon." In *Origin of the Moon,* edited by W. K. Hartmann, R. J. Phillips, and G. J. Taylor. Houston: Lunar & Planetary Institute, 1986.

6: *Captive Moon*

Ashbrook, Joseph. "The Sage of Mare Island." *Sky and Telescope,* October 1962, 193.

Baldwin, Ralph. *The Face of the Moon.* Chicago: University of Chicago Press, 1949.

Brush, Stephen G. "Early History of Selenogony." In *Origin of the Moon,* edited by W. K. Hartmann, R. J. Phillips, and G. J. Taylor. Houston: Lunar & Planetary Institute, 1986.

"Calls Moon Planet Captured by Earth." *New York Times,* June 29, 1909.

Gerstenkorn, Horst. "The Earliest Past of the Earth-Moon System." *Icarus* 11 (1969): 189–207.

Harper, William Rainey. Correspondence. Joseph Regenstein Library, University of Chicago.

Lott, Arnold S. *A Long Line of Ships: Mare Island's Century of Naval Activity in California.* Annapolis, Md.: U.S. Naval Institute, 1954.

MacDonald, Gordon. "Origin of the Moon: Dynamical Considerations." In *The Earth-Moon System,* edited by B. G. Marsden and A. G. W. Cameron. New York: Plenum, 1966.

Mitroff, Ian. *The Subjective Side of Science.* New York: American Elsevier, 1974.

Numbers, Ronald. *Creation by Natural Law: Laplace's Nebular Hypothesis in American Thought.* Seattle: University of Washington Press, 1997.

Osterbrock, Donald. *Yerkes Observatory, 1892–1950: The Birth, Near Death, and Resurrection of a Scientific Research Institution.* Chicago: University of Chicago Press, 1997.

See, Thomas Jefferson Jackson. Correspondence. Mary Lea Shane Archives of the Lick Observatory, University of California at Santa Cruz.

————. *Researches on the Evolution of the Stellar Systems.* Vol. II, *The Capture Theory.* Lynn, Mass.: T. P. Nichols & Son, 1910.

Sherrill, Thomas J. "A Career of Controversy: The Anomaly of T. J. J. See." *Journal for the History of Astronomy* 30, no. 1 (1999): 25–50.

Webb, William L. *Brief Biography and Popular Account of the Unparalleled Discoveries of T. J. J. See.* Lynn, Mass.: T. P. Nichols & Son, 1913.

Wood, John. "Moon over Mauna Loa: A Review of Hypotheses of Formation of Earth's Moon." In *Origin of the Moon,* edited by W. K. Hartmann, R. J. Phillips, and G. J. Taylor. Houston: Lunar & Planetary Institute, 1986.

7: *Sister Moon*

Boussinesq, M. J. "Notice sur la vie et les travaux de M. Édouard Roche." In Édouard Roche, *Mémoires astronomiques et météorologiques,* vol. 1. Montpellier, France: Boehm Fils, 1883.

Brush, Stephen G. *Fruitful Encounters: The Origin of the Solar System and the Moon from Chamberlin to Apollo,* vol. 3. Cambridge: Cambridge University Press, 1996.

————. "A Geologist among Astronomers: The Rise and Fall of the Chamberlin-Moulton Cosmogony," parts 1 and 2. *Journal for the History of Astronomy* 9 (1978): 1–41, 77–104.

Burns, J., J. Lissauer, and A. Mikalkin. "In Memoriam: Victor Sergeyevich Safronov (1917–1999)." *Icarus* 145 (2000): 1–3.

Jeans, J. H. *The Nebular Hypothesis and Modern Cosmogony.* Oxford: Clarendon Press, 1923.

Mitroff, Ian. *The Subjective Side of Science.* New York: American Elsevier, 1974.

Ruskol, E. L. *The Origin of the Earth-Moon System* (in Russian). Moscow: Unified Institute of Physics of the Earth, 1997.

Safronov, V. S. "Sizes of the Largest Bodies Falling onto the Planets during Their Formation." *Soviet Astronomy* 9, no. 6 (1966): 987–991.

Safronov, V. S., and E. L. Ruskol. "Formation and Evolution of Planets." *Astrophysics and Space Science* 212 (1994): 13–22.

"Schmidt, Otto Yul'evich." *http://www.astro.tomsk.ru/post/astronomers/SchmidtOY .html.*

Schultz, Susan F. "Thomas C. Chamberlin: An Intellectual Biography of a Geologist and Educator." Ph.D. diss., University of Wisconsin, 1976.

Tisserand, M. F. "Rapport sur les travaux de M. Roche, professeur de Mathématiques à la Faculté des Sciences de Montpellier." In Édouard Roche, *Mémoires astronomiques et météorologiques,* vol. 1. Montpellier, France: Boehm Fils, 1883.

Wood, John. "Moon over Mauna Loa: A Review of Hypotheses of Formation of Earth's Moon." In *Origin of the Moon,* edited by W. K. Hartmann, R. J. Phillips, and G. J. Taylor. Houston: Lunar & Planetary Institute, 1986.

8: *Renaissance and Controversy*

Alexander, C. M. O'D., A. P. Boss, and R. W. Carson. "The Early Evolution of the Solar System: A Meteoritic Perspective." *Science,* July 6, 2001, 64–68.

Arnold, J. R., J. Bigeleisen, and Clyde A. Hutchison Jr. "Harold Clayton Urey." In *Biographical Memoirs,* vol. 68. Washington, D.C.: National Academy Press, 1995. Reproduced at *http://books.nap.edu/html/biomems/hurey.html.*

Baldwin, Ralph. *The Face of the Moon.* Chicago: University of Chicago Press, 1949.

————. *The Measure of the Moon.* Chicago: University of Chicago Press, 1963.

————. Oral history interview, Niels Bohr Library, American Institute of Physics.

Brush, Stephen G. "From Bump to Clump: Theories of the Origin of the Solar System." In *Space Science Comes of Age: Perspectives in the History of the Space Sciences,* edited by Paul Hanle and Von Del Chamberlain. Washington, D.C.: Smithsonian Institute Press, 1981.

————. *Fruitful Encounters: The Origin of the Solar System and the Moon from Chamberlin to Apollo,* vol. 3. Cambridge: Cambridge University Press, 1996.

————. "Nickel for Your Thoughts: Urey and the Origin of the Moon." *Science* 3 (September 1982): 891–897.

Doel, Ronald. *Solar System Astronomy in America.* New York: Cambridge University Press, 1996.

Home, Roderick W. "The Origin of the Craters: An Eighteenth-Century View." *Journal for the History of Astronomy,* 1972, 1–10.

Jastrow, Robert. "Exploring the Moon." In *Space Science Comes of Age: Perspectives in the History of the Space Sciences,* edited by Paul Hanle and Von Del Chamberlain. Washington, D.C.: Smithsonian Institute Press, 1981.

Krystek, Lee. "The Impossible Rocks that Fell from the Sky." At *http://www.unmuseum.org/rocksky.htm.*

Levy, David. *Shoemaker by Levy: The Man Who Made an Impact.* Princeton, N.J.: Princeton University Press, 2000.

Shoemaker, Gene. Oral history interview, Niels Bohr Library, American Institute of Physics.

Urey, Harold. "The Contending Moons." *Astronautics and Aeronautics* (January 1969): 37–41.

————. Correspondence. Mandeville Special Collections Library, University of California at San Diego.

————. *The Planets.* New Haven, Conn.: Yale University Press, 1952.

————. "The Space Program and Problems of the Origin of the Moon." *Bulletin of the Atomic Scientists* (April 1969): 24–30.

Whitaker, Ewen. *The University of Arizona's Lunar and Planetary Laboratory: Its Founding and Early Years.* Tucson, Ariz.: Lunar & Planetary Laboratory.

Wilhelms, Don. *To A Rocky Moon: A Geologist's History of Lunar Exploration.* Tucson: University of Arizona Press, 1993.

9: *"A Little Science on the Moon"*

Beattie, Donald. *Taking Science to the Moon: Lunar Experiments and the Apollo Program*. Baltimore: Johns Hopkins University Press, 2001.

Brush, Stephen G. *Fruitful Encounters: The Origin of the Solar System and the Moon from Chamberlin to Apollo*, vol. 3. Cambridge: Cambridge University Press, 1996.

Chambers, J. E., and J. J. Lissauer. "A New Dynamical Model for the Lunar Late Heavy Bombardment." Abstract for Lunar and Planetary Science Conference XXXIII (2002).

Cohen, B. A., T. D. Swindle, and D. A. Kring. "Support for the Lunar Cataclysm Hypothesis from Lunar Meteorite Impact Melt Rates." *Science*, December 1, 2000, 1754–1756.

Dones, Luke. Personal interview.

Harland, David. *Exploring the Moon: The Apollo Expeditions*. Chichester, Eng.: Praxis, 1999.

Jones, Eric M., ed. "Apollo Lunar Surface Journal." At *http://history.nasa.gov/alsj*.

Korotev, Randy. "How Do We Know That It's a Rock from the Moon?" At *http://epsc.wustl.edu/admin/resources/moon/howdoweknow.html*.

Levison, H. F., et al. "Could the Lunar 'Late Heavy Bombardment' Have Been Triggered by the Formation of Uranus and Neptune?" *Icarus* 151 (2001): 286–306.

Longhi, John. "The Extent of Early Lunar Differentiation." Abstract for Lunar and Planetary Science Conference XXXIII (2002).

———. Personal interview.

Spudis, Paul D. "What Is the Moon Made Of?" *Science*, September 7, 2001, 1779–1781.

Taylor, G. Jeffrey. "Earth's Moon: Doorway to the Solar System." In Byron Preiss, *The Planets*. New York: Bantam, 1985.

———. "The Scientific Legacy of Apollo." *Scientific American*, July 1994, 40–47.

Weichert, U., et al. "Oxygen Isotopes and the Moon-Forming Giant Impact." *Science*, October 12, 2001, 345–348.

Wilhelms, Don. *To A Rocky Moon: A Geologist's History of Lunar Exploration*. Tucson: University of Arizona Press, 1993.

Wood, John. "Moon over Mauna Loa: A Review of Hypotheses of Formation of Earth's Moon." In *Origin of the Moon*, edited by W. K. Hartmann, R. J. Phillips, and G. J. Taylor. Houston: Lunar & Planetary Institute, 1986.

———. Personal communication.

10: *When Worlds Collide*

Baldwin, R., and D. Wilhelms. "Historical Review of a Long-Overlooked Paper by R. A. Daly concerning the Origin and Early History of the Moon." *Journal of Geophysical Research* 97, no. E3 (March 25, 1992): 3837–3843.

Cameron, A. G. W. "Adventures in Cosmogony." *Annual Review of Astronomy and Astrophysics* 37 (1999): 1–36.

———. Personal interview.

Davis, Donald. Personal interview.

Dutch, Steven. "Velikovsky." At *http://www.uwgb.edu/dutchs/pseudosc/vlkovsky.htm*.

Ellenberger, Leroy. "An Antidote to Velikovskian Delusions." *Skeptic,* 3 no. 4, 1995. At *http://abob.libs.uga.edu/bobk/velidelu.html*.

Hartmann, W. K. Personal interview.

Hartmann, W. K., and Davis, D. L. "Satellite-Sized Planetesimals and Lunar Origin." *Icarus* 24 (1975): 504–515.

Hazen, Robert M. "Reginald Aldworth Daly." At *www.agu.org/inside/awards/daly.html*.

Krystek, Lee. "Venus in the Corner Pocket." At *http://www.unmuseum.org/velikov.htm*.

Levy, David H. "A Marriage of Science and Art." *Sky and Telescope,* June 2001, 78.

Ringwood, A. E. *Origin of the Earth and Moon.* New York: Springer-Verlag, 1979.

Velikovsky, Immanuel. "Before the Day Breaks." At *http://www.varchive.org/bdb/main.htm*.

———. *Worlds in Collision.* New York: Dell, 1972.

Wilhelms, Don. *To A Rocky Moon: A Geologist's History of Lunar Exploration.* Tucson: University of Arizona Press, 1993.

11: *The Kona Consensus*

Alexander, C. M. O'D., A. P. Boss, and R. W. Carlson. "The Early Evolution of the Inner Solar System: A Meteoritic Perspective." *Sciences,* July 6, 2001, 64–68.

Borg, Lars. Personal interview.

Boyarchuk, A. A., E. L. Ruskol, V. S. Safronov, and A. M. Fridman. "The Origin of the Moon: Satellite Swarm or Megaimpact?" *Doklady Akademii Nauk* 361, no. 4 (1998): 481–484.

Cameron, A. G. W. "From Interstellar Gas to the Earth-Moon System." *Meteoritics and Planetary Science* 36 (2001): 9–22.

———. Personal interview.

Canup, R. M., and E. Asphaug. "Origin of the Moon in a Giant Impact near the End of the Earth's Formation." *Nature* 412 (2001): 708–711.

Canup, R. M., and K. Righter, eds. *Origin of the Earth and Moon.* Tucson: University of Arizona Press, 2000.

Chambers, John. Personal communication.

Chui, Glennda. "Huge Scar Unmasked." *San Jose Mercury News,* March 5, 2002, 1F–5F.

Diacu, F., and P. Holmes. *Celestial Encounters: The Origins of Chaos and Stability.* Princeton, N.J.: Princeton University Press, 1996.

Drake, Michael J. "Accretion and Primary Differentiation of the Earth: A Personal Journey." *Geochimica et Cosmochimica Acta* 64, no. 14 (2000): 2363-2370.

————. Personal interview.

Gleick, James. *Chaos: Making a New Science.* New York: Penguin, 1987.

Halliday, A. N., and M. J. Drake. "Colliding Theories." *Science,* March 19, 1999, 1861-1863.

Hartmann, William. Personal interview.

Podosek, Frank A. "A Couple of Uncertain Age." *Science,* March 19, 1999, 1863-1864.

Righter, Kevin. "Core Formation and the Earliest History of the Earth." *Science Progress* 81, no. 1 (1998): 3-16.

————. Personal interview.

Taylor, G. Jeffrey. "Lunar Origin Meeting Favors Impact Theory." *Geotimes* 30, no. 4 (1985): 16-17.

————. Personal interview.

Wilde, S., et al. "Evidence from Detrital Zircons for the Existence of Continental Crust and Oceans on the Earth 4.4 Gyr Ago." *Nature* 409 (2001): 175-178.

Wilhelms, Don. *To a Rocky Moon: A Geologist's History of Lunar Exploration.* Tucson: University of Arizona Press, 1993.

Wood, John. "Moon over Mauna Loa: A Review of Hypotheses of Formation of Earth's Moon." In *Origin of the Moon,* edited by W. K. Hartmann, R. J. Phillips, and G. J. Taylor. Houston: Lunar & Planetary Institute, 1986.

12: *Introducing Theia*

Broad, William J. "Apollo Opened Window on Moon's Violent Birth." *New York Times,* July 20, 1999.

Halliday, Alex N. "Terrestrial Accretion Rates and the Origin of the Moon." *Earth and Planetary Science Letters* 176, no. 1 (2000): 17-30.

Wilford, John Noble. "25 Years Later, Moon Race in Eclipse," *New York Times,* July 17, 1994.

Appendix: Did We Really Go to the Moon?

Federation of American Scientists. "Why the Soviets Never Beat the U.S. to the Moon: Interview with Charles P. Vick." At *http://www.fas.org/spp/eprint/cp_vick_interview.htm.*

Goddard, Ian Williams. "Are Apollo Moon Photos Fake?" At *http://www.badastronomy.com/bad/tv/iangoddard/moon01.htm.*

Godwin, Robert, comp. *Apollo 11: The NASA Mission Reports.* Burlington, Ont., Canada: Apogee, 1999.

Hartmann, William K. "The Paradigm and the Pendulum." *Nature,* April 20, 2000, 817.

————. Personal interview.

Korotev, Randy. "How Do We Know That It's a Rock from the Moon?" At *http://epsc.wustl.edu/admin/resources/moon/howdoweknow.html.*

Lindroos, Marcus, ed. "The Soviet Manned Lunar Program." At *http://www.fas.org/spp/eprint/lindroos_moon1.htm.*

Philips, Tony. "The Great Moon Hoax." At *http://science.nasa.gov/headlines/y2001/ast23feb_2.htm.*

Plait, Phil. *Bad Astronomy: Misconceptions and Misuses Revealed, from Astrology to the Moon Landing "Hoax."* New York: John Wiley & Sons, 2002.

————. "Fox TV and the Apollo Moon Hoax." At *http://www.badastronomy.com/bad/tv/foxapollo.html.*

Redzero (pseudonym). "Moon Hoax Proponents." At *http://www.redzero.demon.co.uk/moonhoax/proponents.html.*

Sibrel, Bart. "A Funny Thing Happened on the Way to the Moon." At *http://www.moonmovie.com.*

Acknowledgments

For me, as for any first-time author, writing this book has been both an exhilarating and a slightly frightening experience. First and foremost, I am grateful to you, the reader, for sharing your time with me. I hope this book has helped you to fall under the Moon's spell (or fall again, if you were a lapsed loony).

I would like to thank John Wilkes, who is the founder and the heart and soul of the Science Communication Program at the University of California at Santa Cruz. Besides giving me the confidence and savvy to prosper in a new career, he taught me to look for the human story that is interwoven with any science story. Thanks also to Peter Radetsky, who helped me believe I could write this book, and to Charles Seife, who incorrectly gave me the impression that one could actually make money as a book writer.

My editor at Wiley, Jeff Golick, was a writer's dream. Thanks, Jeff, for keeping the cap on your editorial pen most of the time, but nevertheless telling me frankly when you got to a section that didn't fit.

Special thanks to the three people who read over my manuscript for accuracy: Joan Cadden (for the science history) and Alastair Cameron and Kevin Righter (for the Apollo and post-Apollo science). They helped me to feel surer of my footing in various bogs where I needed an expert guide. At the same time, I take the responsibility for any errors that might remain.

Finally, I would like to thank my wife, Kay, who is often one of my toughest critics, but at the same time always my staunchest supporter. She went above and beyond the call of duty by preparing the line art for several of the figures in this book, using her wizardry with Adobe Illustrator. On so many levels this book would have been impossible without her.

Index

Page numbers in *italics* indicate illustrations.